grow your
own food

grow your own food

35 WAYS TO GROW VEGETABLES, FRUITS, AND HERBS IN CONTAINERS

deborah schneebeli-morrell

CICO BOOKS

LONDON NEW YORK

This edition published in 2021 by CICO Books
an imprint of Ryland Peters & Small
341 E 116th st, New York, NY 10029
20-21 Jockey's Fields, London WC1R 4BW

www.rylandpeters.com

First published in 2009 as *Organic Crops in Pots*

10 9 8 7 6 5 4 3 2 1

A CIP catalog record for this book is available
from the Library of Congress.

ISBN: 978-1-80065-005-3

Printed in China

Project Editor: Gillian Haslam
Text Editors: Henrietta Heald and
Eleanor Van Zandt
Designers: Roger Hammond and Roger Daniels
Photographer: Heini Schneebeli
Illustrator: Jane Smith

Art Director: Sally Powell
Head of production: Patricia Harrington
Publishing manager: Penny Craig
Publisher: Cindy Richards

Dedication

To my dear friend Jehane Markham, a pioneer
of organic food and a thoughtful gardener.

Author's acknowledgments

Thanks to all my friends, gardeners, and
vegetable growers who shared their gardening
knowledge and who allowed us to photograph
in their gardens. In particular, Mick Rand, Bill
Saunders, Chris Jackson, Alan and Anthea
Stewart, Helen Scott Lidgett, Jill Patchett, and
Gloria Nicol.

I am indebted to Heini Schneebeli for his
perseverance and skill in photographing the
projects in the book during a particularly poor
summer for veg growers.

I would like to thank John Levis of Bygones at
Whitehall Garden Centre in Lacock, Wiltshire for
lending some props and recommend a visit to
his outdoor shop for inspiration and a source of
lovely vintage domestic and garden objects.

Potting mix is commonly referred to as
compost in the UK.

Contents

Introduction

Increasing concern about climate change has made us more conscious of where our food comes from and how it is produced. We know about the unwelcome effects of industrial-scale farming and the negative influence that humans have had on the environment. Sustainability and green issues are at the forefront of many people's minds, and there is a general desire to live healthier lives while lessening our impact on the planet.

We have learned that it is better to eat seasonal vegetables that are locally grown, rather than those that have been flown thousands of miles around the world to reach our stores and markets. As a result, more and more of us are "growing our own." One great advantage of this is that it allows us to harvest crops minutes before eating or cooking them—and to understand fully the meaning of the word "fresh."

Vegetables deteriorate very quickly after they are picked, because the natural sugars start to turn to starch; and by the time commercially grown vegetables reach the stores the flavor is already greatly diminished. That is why frozen peas (which are frozen immediately after harvest) are superior in flavor to so-called fresh peas, which may not reach the stores until a day or two later. Don't just take my word for it. When you start to grow your own vegetables, do a little experiment. Pick and eat a lettuce or carrot or whatever you grow, and then eat a store-bought version, and you will have your proof—there's no comparison.

Pesticide residues, which are still found in commercially grown vegetables, can be harmful to human health, so if you want to be sure that your produce is pesticide-free, you need to cultivate it yourself. Store-bought organic vegetables are relatively expensive— another good reason to grow your own.

Gardeners who grow their own produce are doing real work—the kind of work that originated with our ancient farming ancestors.

Left: *Chilies come in many different varieties and are well suited to being planted in colorful empty olive-oil cans.*

Right: *Borlotti beans quickly climb up tall canes, producing a heavy crop of speckled red pods. These are left to ripen on the vine and are harvested by opening the dried pods.*

Given such a long history, you might be discouraged by the thought that there is too much to learn, too much tradition to absorb; but this knowledge is widely spread, and many people are happy to share their experience, whether by word of mouth or through books and manuals.

Above: *This old, sturdy, wire potato harvesting basket makes a useful container for growing tomatoes.*

It is a fact of nature that plants will grow. Your role is to help to create the right conditions to encourage and enhance that process.

Urban gardening

There is a widespread trend among city dwellers to grow vegetables in the smallest spaces—on windowsills and balconies, in patios and backyards. Unlike many fashions, this one is really welcome. It represents a movement away from the processed and the packaged, away from mass production and shopping—and, importantly, it challenges the domination of supermarket culture. Do we really need particular vegetables and fruit to be available all through the year? Does a strawberry eaten in the middle of winter, grown in a huge industrial greenhouse, bear any relation to an indigenous one grown in the open air and ripened by the sun? Do we want to buy a little packet of plastic-wrapped (cling-film) green beans cultivated in a far-off place and sprayed many times? And do we want to pay high prices for these dubious fruit and vegetables?

More and more people are saying no to these questions. Among them are those of us who do not have much outside space but want to grow our own and enjoy some of the advantages that were traditionally gained from gardening on a vegetable plot or in a larger garden. This is easier than it sounds; many vegetables are adaptable and can be grown in reasonably compact containers.

Many people opt for a grow bag—especially for growing tomatoes. These sausage-shaped plastic bags, filled with potting mix, will fit on a small balcony, but they have little else to recommend them. For one thing, they're extremely ugly; for

another, they're difficult to water. Various gadgets have been designed to help with the watering and to prop up the plants. But the best thing to do with a grow bag is to cut it open and use the mixture in a more attractive container.

Gardening for children

If you have children, encourage them to become involved in cultivating the produce they will eventually eat. This will teach them where and how food is grown, and they will

Above: Chard will grow well in a roomy window box on a sunny windowsill, and adding an attractive shell mulch will help to retain moisture in the potting mixture.

come to realize that treating the earth with care and respect can bring delicious fresh rewards. Sowing seeds and watching them germinate and thrive is a real pleasure, and this small world of growth, harvest, and renewal that you have created will connect you and your family to the vast ecosystem of nature.

Gardening organically

Gardening organically is about working with nature. It is not a scientific process; it is just common sense—and has long been practiced by thoughtful gardeners. If you are growing all your vegetables in pots in a small space, it is not easy to be completely organic, since this would involve making your own compost to feed and enrich the soil. However, if you follow the famous adage "Feed the soil, not the plants," you will be well on the way to creating healthy, sturdy, pest-resistant plants. Where possible, buy organic potting mix.

If you have enough space, make yourself a compost pile; besides reducing the amount of organic waste going to landfill, this will give you a continuous supply of rich compost to add to the soil that you will doubtless have to buy. Avoid pesticides, since these kill beneficial insects as well as those that may be devouring your crops. In reality, aphids are wonderful food for garden birds and for ladybug (ladybird) and syrphid-fly larvae; slugs are loved by frogs and toads; and snails are a delicacy for thrushes—you may have been lucky enough to hear a thrush tapping a snail

Above: *Adding freshly cut comfrey to your potting mix adds a boost of useful nutrients and encourages good cropping of your vegetables.*

shell on an "anvil" stone to reveal the nutritious flesh inside. An organic gardener values and protects these creatures. Feed your crops occasionally, especially while they are producing fruit. Use a natural fertilizer such as seaweed extract or homemade comfrey liquid (see page 16).

The organic method includes companion planting: pairing vegetables with flowering plants that encourage beneficial insects and predators as well as plants with a scent that repels other potential pests (see page 29). It really is a logical system.

Left: *Besides your vegetables, bees will pollinate your flowers.*

Right: *Bees love borage flowers for the rich, sweet nectar they produce. Borage is an annual and grows quickly into a sturdy plant covered in clear blue starlike flowers.*

getting started

The art of growing food crops is neither mysterious nor difficult. Before the widespread proliferation of supermarkets, many people managed to grow a few staple crops in order to feed their family.

We still have the same needs; perhaps we are not so poor, but many of us would like to grow at least some healthful crops to supplement our family diet. And so gardening, particularly vegetable growing, is becoming extremely popular. Gardeners are very generous people and so you will not have to look far to find help and advice. And a huge range of gardening books and magazines is available, from which you can glean much useful information.

If you are intending to grow a few vegetables in containers in a small urban space, it's not difficult to get started. You won't need many tools—a small trowel and fork will be useful, as will a dibble for making holes in which to plant seeds and seedlings.

A few sacks of general-purpose potting mix are essential, and a bag of well-rotted manure is useful to enrich the soil for hungry crops, such as beans and zucchini (courgette).

The nicest task will be finding and recycling suitable containers, and you will find inspiration on the following pages. Some of the projects use mulches to retain the moisture. Besides being practical, mulches can be decorative, if you use collected shells or pebbles.

Perhaps even more fun is the selection of plants. You'll find a dazzling variety of cultivars from which to choose; but remember that, aside from annuals, you need to choose plants that will survive in your hardiness zone. Your local garden center or nursery will offer plants suited to your zone. When buying plants or seeds by mail order, remember to take this into consideration.

Don't worry too much about pests and diseases. These are unlikely to cause much trouble, and your plants will be healthy if you try to be as organic as you can.

Choosing containers

The choice of containers is endless. If you use your imagination and invention to recycle items you already have, you will not only devise interesting planters but also add striking decorative touches to your outside space. Outside space is now often considered an extension of inside space. Just as you might choose a piece of furniture, a vase, or a picture to enhance an interior, you will want to choose pots and other containers to make your exterior space interesting and beautiful.

Once you start looking, you will find all sorts of objects—some discarded, rescued, or borrowed—that can be adapted or converted to make containers for growing produce. Traditional terracotta pots are lovely, but you may prefer the decorative qualities of empty olive-oil cans, for example. If color is your thing, consider acquiring some brightly colored plastic garden buckets. Old baskets can be lined with

Above left: Colored, rubber, all-purpose tubs make excellent, practical, and decorative planters.

Above: This roomy supermarket basket has been planted with spinach, holding enough potting mix to ensure a healthy crop.

plastic to conserve moisture; galvanized buckets and tubs and wooden wine or fruit boxes are among many other options.

The planting projects in this book use a wide variety of containers, each of which has been carefully chosen to match the crop and give it the best chance of success. Some crops need less space than others. For example, quick-growing salad can be sown in a shallow container, such as a dishwashing bowl or a kitchen colander, while rooted crops, including carrots, need more depth. Beans need a long root run; but radishes, being speedy croppers, can thrive in a more limited space.

Above: *Strawberries, both wild and cultivated, grow happily in old wooden fruit or wine boxes.*

Far left: *Arugula (rocket) grows quickly and almost anywhere. All sorts of kitchen containers are well suited to this easy crop.*

Left: *Similarly, radishes are a quick-growing crop and will thrive in these small plastic bowls, and also look pretty.*

Composts and soils

The most important aspect of any gardening, and the key to real success, is to make sure that your soil is fertile, well balanced in essential nutrients, and moisture retentive without being too wet. Enthusiastic beginners often become downhearted at the poor progress of their plants, not realizing that an initial investment of time, thought, and research into soils and potting mixes will provide bounteous rewards.

Start your own compost pile

If you have enough outside space, the best thing to do is to start a compost pile or bin. There are many ready-made closed systems available, most of which consist of a large plastic bin with a close-fitting lid. Among the benefits of a compost pile or bin is a reduction in the amount of waste that needs to be collected from your home. Above all, you will have a rich compost full of beneficial organisms, including worms, to add to your purchased soil. You can buy worms to add to a composting bin. These will devour your vegetable waste and convert it into much sought-after wormcast, a particularly fine and fertile compost.

Potting mix

Don't use homemade compost or pure wormcast for sowing. Vegetable seeds need a sterile soil with no weed seeds or pathogens, which might damage the young seedlings. Poor soil can cause "damping off," in which the base of the new stem rots and the previously healthy seedling flops over and dies.

It is advisable to buy a special seed-starting mix, which will contain a limited quantity of nutrients and fertilizer—you don't want the seedlings to romp away with a lot of lush leafy growth and little strength.

comfrey liquid

Very rich in nutrients, comfrey liquid is believed to have almost magical properties. It is easy to make yourself. Comfrey is a common plant that grows tall and dense; its flowers are much loved by bees. If you have space, you could grow a patch specially but, if not, you may be lucky enough to find some, perhaps in a friend's garden. Cut a plant to the ground – it will quickly regrow. Chop it up and follow the instructions on page 91. Then just add a large splash to a can of water before watering. Tomatoes, peppers, zucchini (courgette) and eggpplant (aubergine) will all benefit from a weekly feed. Comfrey leaves can also be dug into compost or used to make an excellent mulch.

It is possible, with a little searching, to find commercially produced organic seed-starting mix, but check the label. Some manufacturers offer so-called organic products that would not necessarily meet organic accreditation standards. There is no need to be too rigid; just steer clear of chemically enhanced mixtures.

For growing plants in containers, you can choose from a huge range of commercial potting mixes. Or you can make your own mixture, using loam, sand, well-decayed compost, and some damp peat moss. The proportions can be varied, depending on the plant's requirements. For those requiring a rich soil, add some decomposed leaf mold or some well-rotted manure. For improved drainage, where the loam is claylike, some perlite or vermiculite can be incorporated.

An important component of most American potting mixes, peat comes mainly from Canada, which has more than 270 million acres of it. In some other parts of the world, peat supplies are running low, and so horticulturalists there have been experimenting with peat substitutes, such as bark and coir (made from coconut husks). You may wish to try these as a "green" alternative to peat.

Feeding your plants

You can add well-rotted manure to a general-purpose potting mix to make a rich and nutritious medium in which to grow hungry plants, such as beans, peas, and zucchini (courgette). It should not be used for root crops, such as carrots, however, since it will cause the roots to fork or distort. Spent mushroom compost, which contains a lot of straw, is wonderful for lightening the soil, but it is rich in lime, which doesn't suit all crops; strawberries and raspberries, for example, do not thrive in a lime-rich soil.

Above: *Zucchini (courgette) are easy to grow in a large container, but they do need a rich planting mix; add freshly cut comfrey, and mix in a little well-rotted manure or homemade compost.*

Crops grown in containers should ideally be fed a few weeks after planting to boost the fertility of the soil. There are many proprietary organic feeds available—some liquid and some that are incorporated into the surface of the soil. A weekly feed of seaweed extract or comfrey liquid (see opposite) is ideal.

Sowing and growing

One of the gardener's winter pleasures is reading seed catalogs and planning what to grow; but deciding which lettuce, beet (beetroot), or carrot to order can be confusing. When choosing vegetables for pots, look for plants that crop quickly and are described as being suitable for container growing.

Above: *A small trug will hold an assortment of tools, including a trowel and hand pruners, as well as useful odds and ends, such as string, labels, and perhaps a seed catalog.*

Right: *Read the seed catalogs or the back of seed packets so that you find suitable varieties.*

Buying and storing seeds

Organic seeds are readily available from dedicated suppliers, but you may enjoy searching through the seed packets on display at your local garden center. Gardeners collect seed packets throughout the year in the same way that some people collect shoes. Luckily, seeds can be viable for a long time, especially if kept in cool, dry conditions out of the light. Before you buy seeds, read the text on the packet, which will describe the characteristics of the plant, the conditions required for successful growing, and the planting and cropping times. Seeds can be expensive, and the amount of seed varies greatly from packet to packet. You may be able to find some real bargains on the Internet, including seeds from other countries. In particular, some Italian seed producers offer high-quality seeds, including vegetables, herbs, and flowers, at very reasonable prices.

Growing from seed

Seeds are programmed to germinate and grow into mature plants. To thrive, they need warmth, light, and moisture, as well as a sterile growing medium—never use garden soil or recycled planting mix, because these will contain fungi and bacteria that could damage emerging seedlings. There are various seed-starting mixtures on the market; choose one of these for best results. Don't let the seedlings dry out, and never expose them to extremes of temperature.

If you are growing a small number of vegetables, you can sow them in trays or pots and let them germinate on a sunny windowsill. It is a good idea to grow tomatoes, chilies, and peppers in this way, because they need to be started early in the year, in a warm place, in order to have a long enough season to produce a good crop.

Most keen gardeners want a greenhouse. Apart from its practical advantages, a greenhouse can be a wonderful haven from a busy world, a place to be alone with your thoughts. It may be that you can only aspire to such a luxury, but there are many other ways of sowing seeds and nurturing seedlings. One of the most inventive is to create a mini greenhouse or cloche from a large clear plastic storage box with a detachable lid. If this is placed in your outdoor space with the lid on, the internal temperature will be a few degrees higher than outside. Remember to ventilate the box on warm days by putting a stick under the edge of the lid; otherwise, the tender young seedlings you have planted could be damaged by a sudden buildup of heat.

Left: *Using clear plastic storage boxes as a cold frame is a good idea. With the lid in place, you will be able to increase the temperature inside the box. This is important, since warmth is what makes seeds germinate.*

Above: *After you have sowed your bean seeds, you could speed up germination by covering the potting mix with a piece of glass. (For safety, make sure it has smooth edges.)*

All kinds of container are suitable for sowing seeds, including plastic trays, pots, cells and modules, egg boxes, and even supermarket packaging. A plastic container with a lid—the kind that bush fruits are sold in—can be used as a mini cloche; such containers are usually perforated in the base and the lid, providing a convenient means of drainage and ventilation.

Clear plastic drinks bottles can be adapted to drip-feed plants, to make a temporary cloche, or to protect vulnerable young plants against snails and slugs.

The cardboard tubes inside toilet-tissue rolls or paper towels make good biodegradable pots, especially for larger seeds, such as peas or beans. Alternatively, you can make paper pots from strips of newspaper, using a clever little tool designed for this purpose (see page 24).

If you sow seeds sparingly, you will be able to minimize disturbance of the seedlings when

Left: *A cut-down plastic water bottle makes a useful collar around a young plant, protecting it from the hungry advances of slugs and snails.*

Below: *Some seeds, such as basil, seem to have 100 percent germination, so you will need to thin them out and plant them apart so that mature plants have enough space to grow.*

thinning later. It is sometimes a good idea when sowing fine seeds—carrots in particular—to mix them with a little sand, because this helps to disperse the seeds. For larger seeds, such as zucchini (courgette), peas, cucumbers, or beans, sow two in each pot or cell and discard the weaker when they have both grown their first set of real leaves.

The next stage is pricking out—selecting the healthiest seedlings and replanting them, more widely spaced, in new potting mix (see page 16).

Hardening off is the important process that occurs between the early protected seedling growth and the permanent transfer outside of the sturdier mature plants. To achieve this successfully, leave the lid off the cloche during the daytime for a few days until the plants become adapted to cooler conditions. Eventually, when the nights are warmer and any danger of frost has passed, take off the lid or take the plants outside permanently.

Above: *Use the space around potted trees to plant a crop of salad, scallions (spring onion), or arugula (rocket).*

Intercropping

Intercropping is a method of planting a smaller, fast-growing crop using the space between slower-growing larger vegetables. The quick crop matures just as the slower crop begins to need more space. Typical combinations include lettuce grown around cabbages and broccoli, and radishes and scallions (spring onion) grown between carrots.

Decorative standard trees are often grown in pots (olive trees are particularly popular). Instead of leaving bare soil at the base, use it to plant a salad crop—lettuces or scallions (spring onion) will thrive in these conditions, and even larger onions will grow well. Arugula (rocket) and radishes will be happy, as will the large variety of cut-and-come-again salads. Dig a little organic fertilizer into the mixture at the top of the pot before planting your intercrop.

Buying young plants

If you don't have the time or space to grow all your vegetables from seed, you can buy young plants. Good nurseries will provide healthy ones ready to plant out. It is a sensible idea to buy tomatoes in this way, because you can order a selection of different varieties to grow together. Chilies, eggplants (aubergine), and cucumbers are also usually available as young plants and are normally planted out from late spring, when the danger of overnight frost has passed.

Plant support

Some plants need a structure to climb or scramble over. Bamboo canes are the usual choice, although in Britain "pea sticks," cut from various trees (see page 116), are a popular choice. The height of the canes or sticks will depend on the variety of pea; they vary from dwarf size to 6 feet tall (1.8m). Pole beans, or scarlet runner beans, grow very tall and need to twine around rods or canes. Dwarf green beans may need tying in to stop from flopping over the sides of the container. Beans are easier to pick if well supported.

Newspaper pots

The best kind of pots are those that can be planted directly into the soil and allowed to degrade naturally, such as the newspaper pots shown here. They cut down on work, since you don't have to remove the plants from the pots, and the seedlings are healthier because their roots are not disturbed in the way they would be when potting on from a small to a large container. The green advantage to this method is that it removes the need for plastic pots.

There are various ways of making paper pots (see the Internet), but one of the simplest uses a two-part tool called a Paper Potter. This consists of a wooden cylinder, which is pushed firmly into a wooden base—an action that folds and secures the paper to form a pot. Paper pots are ideal for planting peas and beans of all kinds, because these plants have long roots that will penetrate the paper.

Top: *You will need to buy in strawberry plants early in the season if you want them to bear fruit the same year.*

Above: *Unless you have enough space to grow chilies from seed much earlier in the season, it is best to buy them as young plants. Look for those grown organically.*

Above: *To use a Paper Potter: Tear the newspaper into strips about 4 inches (10cm) wide and 20 inches (50cm) long. Wrap a strip of newspaper around the potter a few times so that it reachesjust over the cylinder at the handle. Overlap the base by about 1¼ inches (3cm) . As you push the cylinder firmly into the base of the potter, the paper will fold to make the bottom of the pot. Remove the cylinder from the base and carefully pull it out of the pot. Fill the pot with the potting mix in preparation for planting.*

Watering and mulching

Vigilance about watering is vital when growing plants in containers. Only a limited amount of moisture can be stored in a pot, and most pots lose water through evaporation, mainly from the surface of the potting mix. A potentially leaky container, such as a basket, should be lined with plastic. Don't forget to make drainage holes in the base; waterlogging is as damaging as drying out. To enhance drainage, put a layer of broken pots, or crocks, in the base of your container before adding the potting mix; stones, pebbles, grit, sharp sand, and even broken polystyrene containers are suitable alternatives.

Various products are designed to aid water retention in compost; these are usually in the form of granules that swell and take up water. They must be used carefully and only in small containers. Although one or two projects in this book include water-retaining granules, I am not generally in favor of them, because they alter a natural process and can lead to overwatering.

When to water?

It is preferable to check the potting mix regularly, looking for signs of thirst in your plants and giving plenty of time and thought to the process of watering. Early morning or later in the evening are the best times to water. Plants can be damaged by watering in the hot sun; water droplets on a plant act like a lens, allowing the sun to burn the leaves. Too light a watering at the wrong time will encourage the roots to reach for the surface and risk scorching. I regard watering as my quiet, observant, thinking time—the special time that I spend on my own with my plants.

Mulches

An effective way to counteract evaporation is to add a mulch around the base of the plants on the surface of the soil. Almost anything will do. Bark chips make a dark covering, whereas broken china or seashells are more decorative; comfrey leaves are nutritious as well as attractive. If you are growing strawberries, straw has the added advantage of lifting the berries off the soil and away from slugs.

Color and tone of mulches can be important—for example, the light tone of crushed seashells spread around the base of rainbow chard will reflect the light and speed up their growth.

Left: *Mulches are practical as well as decorative and can include crushed seashells, pea gravel, and slate pieces. These are available from garden centers and suppliers.*

Right: *The idea is to cover the exposed soil, so a variety of materials will do—try straw, polished pebbles, chipped bark, china shards, and terracotta crocks.*

Dealing with diseases and pests

A great fuss is made about plant diseases and pests, much of it generated by chemical companies seeking to sell products designed to deal with these problems. Commercial preparations are usually best avoided, although there are a few benign chemicals that are permitted in the organic system.

In reality, the most effective way to guard against any disease or pest is to grow sturdy, healthy plants in a well-balanced potting mix. Don't overcrowd your plants, and allow good air circulation. Make sure they have enough light; sun is important, but don't let them overheat. Water regularly, and go easy on the feeding—never give more than the amount specified on the packet.

There is growing opposition to the use of chemicals in horticulture because of the damage they can do to the environment. The good news is that there are not many serious pests to contend with if you are cultivating vegetables in the controlled environment of a container.

Insects

Although many insects, notably bees, are beneficial in the garden, a number of species can do significant damage to fruit and vegetables. Aphids are among the worst offenders. They attack a wide range of plants, including beets (beetrot), eggplant (aubergine), peas, squash, and tomatoes. They can damage a plant by sucking the sap; the telltale signs are twisted or distorted leaves. The juicy bugs make nutritious food for fledgling birds; syrphid-fly and ladybug (ladybird) larvae also rely on a healthy diet of aphids. If infestation is causing damage, a good blast of water from a hose will dislodge most of them.

Above: *A small snail, so often the gardener's enemy, is fascinating to watch as it glides on its slime trail across a leaf. Collect them at night by flashlight.*

The maggot of the root fly burrows deeply into carrots, while the flea beetle nips tiny holes in radishes, arugula (rocket), some oriental salads, and brassicas. Fortunately, these two insects cannot fly above a height of about 8 inches, so can be deterred by erecting a low barrier—made of nylon mesh, for example—around the plants.

Slugs and snails

Slugs and snails are greedy feeders, attracted by soft new growth, and they can demolish a plant overnight. All gardeners hate these pests, but far too many toxic pellets are used to control them, with the result that the soil is contaminated and the poisoned carcases of the snails and slugs may be eaten by toads or thrushes, with devastating consequences for those beneficial creatures.

There is an organically approved pellet made from ferrous sulfate, which breaks down in the soil, but the best method of dealing with slugs and snails is to be vigilant and pick them off plants and containers when you spot them.

Slugs and snails are dormant during the day, sticking under the rims of pots or shady places out of the light. They feed at night, so a visit with a flashlight can result in quite a harvest. Collect them in a bucket of salt water, which will kill them quickly.

Another option is to place a mechanical barrier around a tender seedling—a plastic water bottle with the base cut away works well. An extra precaution is to wrap a strip of copper around the top of a container; slugs and snails won't cross the metal.

Scooped-out grapefruit shells placed skin up around your vegetable plants make handy hiding places for slugs and snails. Simply check them each day, and remove the culprits.

Cats, birds, raccoons

If gardening on a patio, roof terrace, or windowsill, you won't be challenged by rabbits, deer, or mice, but you may be affected by cats; cats love fresh soil and may think you have provided them with a convenient cat tray. Until your plants have grown, you could use a hard mulch, such as china shards or shells, to deter them.

Birds may be a problem. In towns, pigeons can do a lot of damage; they love to peck at peas, chard, and cabbages. Blackbirds like berries of most kinds. Anything that flaps or whirrs—such as a child's pinwheel—will scare birds off. Alternatively, try a few old CDs suspended on strings; the flashing of the silver side is a deterrent.

A raccoon will regard your cherished crops as its special buffet. If your plants are on a balcony, they're probably safe, but if the containers are positioned on a patio, you may have a problem and need to resort to a wire-mesh fence to protect them from the raccoon's attentions—preferably a wire-mesh fence that's floppy along the top.

Companion planting

Grow your vegetables alongside other flowering plants if you can. The companion flowers will attract pollinating insects, which will also visit your bean and tomato flowers, your zucchini (courgette), peas, chilies, peppers, and eggplants (aubergine). Chives or garlic alongside carrots will deter the root fly, and the pungent aroma of basil or French marigolds growing near tomatoes will discourage white aphids.

The herbs savory and thyme grown around fava beans will deter infestation of black aphids. Stately fennel and borage, with its lovely star-shaped blue flowers, both invite bees and hover flies, the two most beneficial insects in the vegetable garden.

Above: *By far the most useful and well tested companion plants are the fiery orange marigolds (Tagetes), with their pungent scent that deters aphids. They have the added bonus of having a long flowering season.*

essential herbs

Herbs can be grown successfully in the smallest of outside spaces, even a small balcony, or in a kitchen. As long as they have good soil, plenty of light, and a fair amount of sun and warmth, most herbs will thrive in containers. If you are new to vegetable gardening, it is no bad thing to start by growing herbs. You will soon graduate to growing vegetables, and then you can put your herbs and vegetables together in the kitchen to make delicious meals.

Among hardy perennial herbs that can survive at quite low temperatures are the alliums (hardy to zone 4). This family includes chives, a useful everyday herb that is easy to grow and has pretty purple, edible flowers. Welsh onions are similar to chives but have rather bulbous hollow leaves; and garlic chives have straplike leaves and lovely white flowers.

Sweet majoram, oregano, savory, and thyme will tolerate dry conditions. The taller, untidy, tarragon needs more space, but is also worth growing—make sure you choose the French type, rather than the Russian, since the latter has a rather weak flavor. Rosemary and sage are larger evergreen shrubs with a woody framework; both should be planted in generously sized pots with a good depth of potting mix. Many varieties of mint are also popular with gardeners—Korean and American mountain mints are good for making tea.

Annual herbs, which last for one season only, should be grown from seed. Indispensable varieties include basil, dill, and cilantro (coriander). Garden chervil, which grows in a pretty mound of feathery leaves, is another annual herb, and is useful for cooking. But don't confuse garden chervil (*Anthriscus cerefolium*) with wild chervil (*A. sylvestris*), an invasive species that hosts a viral disease infecting some vegetables. Annuals will flower and set seed in a season.

Your herb garden should, of course, include parsley. This is a biennial, which means that it flowers and sets seed in the second season after planting.

A basket of mixed herbs

Traditional cane baskets are still widely available and generally good value; examples from the 1950s often have a decorative strip of plastic threaded around the rim and handle. A well-loved vintage basket that is slightly worn and can no longer hold shopping securely is ripe for recycling as an original planter. It must be able to hold enough potting mix to create a good growing environment. You will need to line it—first with plastic to protect the natural materials in the basket, then with felt (or something similar.) Adding water-retaining granules will help to keep the mixture moist and reduce the need for endless watering on hot days. A bark mulch also reduces evaporation from the surface of the mix.

green*care*

Thyme (lots of different varieties, but especially lemon thyme), flat-leafed parsley, marjoram, and chives are all particularly well suited to container growing.

Right and far right: *You may like to buy small herb plants to pot on in a larger container, as here. They are generally inexpensive, but it can be more rewarding to grow your herbs from seed or cuttings.*

1 Push the plastic bag into the basket so that it covers the whole of the inside, coming well up the sides of the basket. Cut a few slits in the base to allow for drainage.

2 Place the liner on top of the plastic and push it into place, trimming the edges to fit the shape of the basket, if necessary.

3 Add the potting mixture, mixing in a small amount of the water-retaining granules (follow the instructions on the package) in the lower section of the mixture.

4 Plant the herbs, placing the thymes at the front and the taller parsley and marjoram at the back. Water well and cover the mixture around the herbs with bark mulch.

herb hints

- Start by growing the most familiar culinary herbs. Then experiment with a few more unusual types and find new recipes that incorporate them.

- Basil is an indispensable container herb (see pages 38 and 42). Every one of the tiny black seeds will germinate. A few small basil seedlings, ready to pot on, make an excellent present for a keen cook.

- Garden chervil, my favorite, is a less well-known herb with an aniseed flavor (see page 36). It is delicious in salads and on buttered new potatoes.

- Fennel, with its graceful habit and feathery leaves in green or bronze, must have a deep container to accommodate its long roots.

- Bushy perennials, such as rosemary and sage, need large pots filled with a potting mix that is more soil-based than that used for other herbs.

- Dill, cilantro (coriander), and summer savory thrive in pots. To keep them healthy and productive, give them plenty of water and harvest often.

- Mints are rampant growers and need to be confined in containers (see page 45); otherwise they will quickly overgrow everything in their path. Repot mints and replenish the soil regularly to retain the herbs' depth of flavor.

Above: *Cilantro is a quick-growing herb often used in Middle Eastern and Asian cooking. Pick the leaves to prevent the plant flowering (although the flowers are edible).*

Right: *The leaves of African basil may be used in cooking. They have a mellow flavor and the scent of cloves.*

Far right: *Purple sage leaves are delicately colored.*

Charming chervil

Garden chervil is a pretty herb that deserves to be better known and more widely grown. It has delicate fernlike leaves and white umbelliferous flowers resembling those of parsley. The older leaves turn subtle shades of pink, especially in cool weather.

Chervil is widely used in French cooking, in salads, soups, sauces, and omelets. It is easy to grow from seed and will thrive in cool, damp conditions in partial shade. Although normally treated as an annual, garden chervil will survive the winter in a mild climate—worth a try if you live along the Pacific Northwest coast, for example—and it's a vigorous self-seeder. Although you can sow the seed in spring, I find it grows better when sown from midsummer onward.

green*care*

If you sow chervil seeds in a wet period, watch out for slugs and snails, which enjoy eating the young seedlings. See page 28 for ideas for keeping them at bay.

Far right: *A wicker basket makes a suitable container for garden chervil. Remember to line it with plastic to retain moisture.*

Right: *Garden chervil has delicate lacy leaves. It will tolerate a certain amount of shade and, in a mild climate, it may even continue to grow in winter. Light and temperature affect the coloring of the leaves—the occasional pink or purple ones look lovely in a salad. For a constant supply of chervil, sow seed every few weeks from midsummer.*

Sweet basil in a clay pot

you will need

seeds of sweet basil

seed-starting potting mix

seed tray

clay pot

mixture of a loam-based and a soilless, peat-moss-free potting mix

dibble

One herb I could not live without is basil. Its clove and aniseed scent is surely the most penetrating fragrance of all the herbs. A perfect partner for succulent homegrown tomatoes, it is also the key ingredient of pesto—the most exquisite sauce ever created. To make pesto, simply pound fresh basil leaves with extra-virgin olive oil and combine with crushed pine nuts, grated Parmesan cheese, and chopped garlic.

Basil plants are easy to grow and will thrive inside or out as long as they are in a warm, sunny position. In Mediterranean countries, you will see them on windowsills and doorsteps, sometimes planted in empty olive-oil cans. Each variety has a subtly different flavor.

The most commonly grown variety, sweet basil (*Ocimum basilicum*) is a reliable half-hardy annual, meaning that you sow the seeds each year and the plant dies down in the fall. The blowsy aromatic leaves are large and crinkled, releasing their pungent aroma at the gentlest touch. Thai basil has smaller leaves, with purple undersides, and is widely used in Thai sauces and curries; it has a more pronounced aniseed flavor and is always cooked. Bush basil, or Greek basil, has smaller leaves and makes a decorative, compact plant. Its flavor is less intense than that of sweet basil.

green*care*

Sturdy basil plants grown from seed, planted in rich potting mix, and nurtured by the summer sun are in a different league from the sappy supermarket pots of overcrowded seedlings, which will last a few days only. Homegrown basil should last the summer. To extend its life, pick off any flowering shoots.

Right: A generous traditional terracotta pot containing a few well-spaced plants will provide you with enough fragrant basil leaves to last the whole summer.

1 Read the seed packet carefully, and follow the sowing instructions. Fill the seed tray with the seed-starting mix and pat down. Sprinkle the seed thinly over the surface. Add a fine layer of the mix until the seeds are covered. Water with a fine rose. Keep the seeds in a warm place, such as on a sunny windowsill, until they germinate.

2 Wash out the clay pot thoroughly, and place a few large stones or broken crocks in the base to encourage good drainage.

3 Fill the pot with the combined potting mixes and tamp down. Use a dibble to make holes in this large enough to take the seedlings without squashing their roots.

4 Carefully lift the seedlings out of the seed tray, first loosening them with the dibble. Hold the seedlings by the first leaves only, to avoid squeezing or damaging the delicate stalks.

5 Space the seedlings a few inches apart—the plants will grow large, and if there are too many, they will be vying for the food in the soil. Water the seedlings well, and keep them out of the hot sun for a few days until the roots become established in their new position. Water regularly and wait patiently for the plants to grow.

Right: *"Purple Ruffles" is an extravagant variety of sweet basil that has deep purple, crinkled leaves. It is less hardy than the green variety, and its leaves are less aromatic, but they are used torn in salads to add spectacular and unusual color.*

African basil and sage

Both these two bushy shrubs grow well in containers, providing the pots are big enough to hold a good depth of potting mix. With African basil especially, bear in mind that the pot may have to be moved in the winter, because this plant doesn't much like cold weather.

Right: *African basil is a shrubby perennial, meaning that it does not set seed or die back. It has decorative, blue-tinted leaves with purple veining, and long spikes of lilac-colored flowers, which are irresistible to bees.*

Above: *Sage, like rosemary, is an evergreen shrub from the Mediterranean area. It has a pungent flavor that goes well with liver, potatoes, and squash. Varieties include a variegated sage with cream and white blotches on the furry gray leaves, purple, common, and broad-leaved sage. They all bear attractive bluish purple flowers, which are very attractive to bees. Keep the bush trimmed, and transfer into a larger container after a year or so if you want the plant to increase in size.*

herbs in rose pots

If you are interested in food and cooking, it is essential to have a good supply of herbs within easy reach of the kitchen. The herbs chosen for these tall rose pots (also called long toms) are some of the most widely used in Mediterranean countries.

- Rosemary (seen in the large pot on the left of the photo) makes a woody evergreen shrub, which can be clipped into a topiary shape. It is often added to lamb or chicken before roasting and makes a wonderful partner to garlic when added to a skillet of sautéed potatoes.

- There are so many varieties of thyme that it is hard to know which one to choose. A pale-leaved common thyme forms an attractive miniature shrub (seen in the large pot on the right). The smaller pot on the right contains a creeping thyme that will have pretty pink flowers. Lemon thyme, a chefs' favorite, is also well worth growing. The secret with thyme plants is to keep them well trimmed and not to let them grow leggy and untidy. I like to sprinkle a few sprigs of thyme along with some coarse salt and good-quality oil on a butternut squash, cut in half lengthwise, before baking in the oven.

- Golden majoram (seen in the small pot on the left) is a spreading, bushy herb that can be cut back each spring to regenerate. Add it to beet (beetroot) and feta salad with extra-virgin olive oil and fresh lemon juice.

- All these Mediterranean herbs will grow well for a year or so in the rose pots, but, as they enlarge and want to grow into shrubby plants or bushes, they should be transplanted to larger containers. They love to grow in sun and will tolerate some drought. Plant in a rich loam-based potting mix.

Below: *If you are short of space, plant your herbs in rose pots. Since they are tall, they hold more potting mix without using too much horizontal space.*

Galvanized buckets filled with mint

Above: *The pale mauve flower spires of black-stemmed mint, loved by bees, appear late in the season. Mint leaves lose some of their pungency when the plant flowers, so cut off the blooms if you want to use the leaves in cooking. Otherwise, crop strongly flavored mints, such as spearmint and Moroccan mint, and leave other varieties to flower.*

Left: *Plant a selection of mints in vintage galvanized buckets. You just need to make a few drainage holes in the base to make a perfect planter.*

Mint is widely used medicinally (as a digestive and decongestant), in cosmetics (as a cleanser and stimulant), in the kitchen (as a flavoring), and to make teas. There are hundreds of varieties, offering many different flavors, so before growing mints for use in drinks or food, compare mature plants by rubbing the leaves between your fingers and choose the ones that have the scent you prefer. American mountain mint is perfect for making a refreshing tea, while apple mint is often used to flavor the water when boiling new potatoes. Pennyroyal, a powerful, low-growing peppermint, makes a useful spreading plant along the edges of paths and patios. As you brush past or tread on the leaves, the intense aroma will fill the air. I rub it into my face and hair to stop the midges from biting when I am working in the garden late in the evening.

It makes sense to grow mint in a pot sunk into the soil, which prevents the vigorous roots from overrunning nearby plants. The herb likes some shade and plenty of moisture. Mint grown in a container will do well for a year. It develops into a large plant very quickly, using up all the nutrients in the soil. This means that the intensity of the flavor will be reduced; to restore it, replenish the soil regularly each spring—or consider starting again with a new young plant.

These vintage galvanized buckets are inexpensive and make ideal containers for a variety of contrasting mints. Keep them by your back door so you can easily pick a few leaves when needed.

growing mint

When you remove a young mint from its original pot for planting in your own container, you will see how the roots are raring to go—they will take off as soon as they reach some lovely new soil. You will probably have to divide the vigorous plants in spring and replenish the potting mix.

Mint can also be used as a companion plant (see page 29) to repel insects, but its vigorous growth means that you do need to be careful where you plant it.

Welsh onions and garlic chives

Chives, garlic chives, and Welsh onions (also called Japanese bunching onions) belong to the allium family, which also includes garlic and leeks. Ordinary chives may start to look rather tired as the season progresses, especially if they are grown in a container; it is worth cutting the leaves back to encourage a healthy regrowth. Garlic chives have straplike leaves with a mild garlic flavor and are much favored in Chinese cooking. The pretty white flowers are edible and, pleasingly enough, bloom late in the year. All in all, this is a rewarding and tidy plant, which, because of its good looks, is often grown in the flower border.

Welsh onions are hardy (to zone 0) plants that can grow quite tall. Their tubular leaves, attached to a bulbous thickened base, may remain green all year; they have a strong onion flavor.

green*care*

You can grow chives, garlic chives, and Welsh onions from seed, but to save time, you may prefer to buy them from a nursery or garden center as young plants. They will enlarge quickly, creating more bulbs, which will make the plant congested in the pot in time. However, alliums are easy plants to propagate. Simply divide the clumps by pulling apart the bulbs, and create several new plants by planting a group of them in a separate pot.

Left: *The fluffy white globe flowers of Welsh onions sit on top of the fat, hollow stems and, like the more finely cut purple chives, can be very decorative. Use like chives.*

Right: *The flowers of garlic chives are very welcome at the end of the summer when other plants are dying down. They look good planted in this unusual ribbed clay pipe. Place it on a clay saucer or directly on the soil.*

leaves and shoots

My aim is to cut some green leaves to eat every day of the year. To achieve a year-round harvest, you need to sow seed successionally—that is, sow more of the same seed every few weeks. In winter, protect your plants from cold, wet weather by covering them with a coldframe or a cloche.

Salads are quick and fairly easy to grow. Since they are shallow-rooting, they will thrive in containers. Window boxes, dishwashing bowls, buckets, and all sorts of domestic containers are perfect. Plastic containers are particularly good because they allow minimal evaporation. Water your salad plants regularly in the evening, and don't let the soil dry out; prolonged drought will stunt the growth of a salad plant and produce leaves that are tough and bitter. For a colorful salad, grow a mixture of red, green, speckled, crinkly, and smooth leaves—each variety has a different taste.

Another must is peppery-flavored arugula (rocket). Both the wild variety, with its finely cut leaves, and the cultivated type, with its nutty, edible flowers, are quick to germinate and grow. Oriental greens—the red-veined mustard and cut-leaved mizuna, and corn salad (also known as lamb's lettuce)—do better when sown later in the summer; they seem to respond to the cooler fall weather and shorter days. If you sow seed several times, at regular intervals, you will be able to harvest leaves until late in the year.

Chard is another leaf crop that is decorative as well as being a versatile vegetable. Grow broad white-stemmed Swiss chard or varieties with more unusual colors. 'Bright Lights', with its lemon-yellow, pink, and white ribs, and the exotic fuchsia pink ruby chard are quite magical.

Beets (beetroot) and carrots look wonderful when their vibrant hues contrast with a colorful container. Radishes grow well in a small container but will not thrive in hot sun. Keep them moist and cool, and sow every two weeks to ensure a constant supply.

Wild arugula (rocket) in a kitchen colander

green*care*

Start to harvest the arugula (rocket) when each plant produces a number of strong leaves. Cut them as you need them, and watch them regrow. After a number of cuttings, the plant will want to flower. At this point, your second crop, sown a couple of weeks later than the first, will be reaching maturity.

Arugula (rocket) is easy to grow, which makes it a popular crop in the vegetable garden. Quick to germinate and mature, it can be sown successively throughout the summer to ensure a continuing supply. Alternatively, sow once or twice and treat as a cut-and-come-again crop, cutting just a few leaves at a time from each plant and then leaving it to produce more.

Wild arugula (rocket) has finer cut leaves and a more peppery taste than the cultivated variety. It is a perennial and can be overwintered because it is able to withstand frosts. Cultivated arugula (rocket) is an annual with tender leaves and a relatively mild taste. When it eventually runs to seed, it produces beautiful cream-colored, four-petaled flowers.

If you want a supply of arugula (rocket) throughout the year, it will grow happily on a kitchen windowsill during the cold months. Keep your arugula (rocket) plants well watered, especially in hot weather, when they have a tendency to bolt or run to seed. If they suffer from a lack of water, the leaves will become tough and unpleasant to eat.

One pest that may attack arugula (rocket) is the flea beetle, which bites tiny holes in the leaves. The beetle is unlikely to trouble container-grown plants, as it cannot jump high enough to reach most pots, but if it is a problem, you can cover the crop with a fine mesh or net.

When it comes to using recycled containers in your vegetable garden, aluminum kitchen pots, pans, sieves, and colanders are undiscovered classics, which are widely available at rummage sales, flea markets, and thrift stores. The old aluminum colander used here resembles a garden urn and makes an entertaining but practical container for arugula (rocket). It is large enough to hold enough potting mix for the crop, although a water-retaining mat has been added in the base to conserve as much moisture as possible.

Right: *This old colander makes a perfect urnlike planter for the finely cut leaves of wild arugula (rocket). You will easily find similar examples in a rummage sale or secondhand store. Add a few old aluminum pots and pans, and you can make a real "kitchen garden."*

1 Put the hanging-basket liner inside the colander—you may need to cut and overlap to fit if the sides of the colander are steeper than those of a hanging basket.

2 Put a little potting mix into the base of the colander and cover it with the water-retaining mat. There's no need to cut this to fit the base of the container.

3 Fill the colander up to the rim with potting mix, press down firmly, and sow the seeds thinly over the surface of the mixture. Sprinkle a fine layer of mixture over the seeds to cover them, pat down, and water well with a fine rose. (It is important to use a fine rose, because a jet of water would wash all the seeds to one side.)

4 Keep the potting mixture moist at all times. When the seeds have germinated and the seedlings are growing strongly, you will need to thin them out to allow each plant sufficient space to grow. If you are very patient by nature, you could even wash all the surplus seedlings and add them to a mixed salad.

Right: *The flowers of cultivated arugula (rocket) are edible, and add a nutty flavor to salads; they are superior in taste to the yellow flowers of wild arugula (rocket).*

Cut-and-come-again salad leaves

Growing cut-and-come-again salad leaves makes it possible to have fresh salad ingredients at your fingertips every day throughout the summer. For a continuous supply, you need to sow successively—that is, every couple of weeks or so. These simple, inexpensive plastic window boxes are easy to find in markets and stores.

"Cut-and-come-again" means just what it says. Instead of growing individual lettuces for harvesting as a whole, you grow a number closer together and crop individual leaves as they grow. The idea is that the plant will then grow new leaves, which can be continually cropped over a few weeks. If you sow seeds a couple of weeks apart through the season, you can extend the harvest over several months. You can even have a late crop when the weather is colder, as long as you cover the plants with a cloche or fleece.

Baby spinach leaves can also be grown as a cut-and-come-again salad crop, as can mizuna, mustard, and corn salad (or lamb's lettuce). If you sow these later in the season—ideally, after midsummer—they seem to do better and suffer less from problems such as bolting. Sow the seeds directly in the prepared potting mix, and thin out when they are 1–2 inches high, allowing at least 4 inches between plants. You can eat the sweet-tasting thinnings in a salad or use as a garnish.

you will need

brightly colored plastic window boxes

multipurpose peat-free organic potting mix combined with a little loam-based mixture

seed of spinach

seed of mixed salad leaves

polystyrene planting tray, broken into small pieces

green*care*

Seed dealers sell a wide variety of cut-and-come-again salad crops, and if you sow over a season, you will be able to try out quite a few. Mixtures are attractive, often combining red and green lettuces as well as plain and curly leaves.

1 Put the broken polystyrene pieces in the base of each window box. This layer of drainage material should be 1–2 inches (2.5-5cm) deep.

2 Fill the window boxes almost up to the rim with the potting mix. Pat the surface with the palm of your hand to make it even and level.

3 Put the spinach seed in the palm of one hand, and use the other to sprinkle it lightly onto the potting mix.

4 Gently sprinkle a light layer of potting mix over the seed so that it is all covered.

green*care*

If you sow your seed early in the season, you can speed up germination by placing a sheet of glass or plastic or even a large clear plastic bag over the window box. This creates a mini greenhouse and increases the soil temperature.

Right: *These brightly colored plastic window boxes are very inexpensive. They are light and easy to move and make a brilliant contrast to the pure green young leaves of spinach, as well as the decorative curly leaves of the red lettuce. Pick the leaves when young and tender and water well. Lettuce will quickly become tough and bitter if short of water.*

5 Repeat with the other box, sowing salad leaf seed. This is much finer seed, so sprinkle it lightly. Water both boxes with a fine rose, so as not to dislodge the seeds.

6 When the seedlings are 1–2 inches (2.5-5cm) tall, thin them out to allow the remaining ones to grow sturdier. Allow 2–4 inches (5-10cm) between each seedling.

Red lettuce and shiso in enamel tins

When you grow vegetables in a small space, you don't want all your crops to be green. Luckily, salad vegetables also come in many tones of purple and red—from radicchio, with its deep-red veins, to the vibrant magenta stems of ruby chard and the purple globes and leaves of some of the older varieties of beet (beetroot), while the fuchsia pink

of radish flashing against deep-brown soil is both intriguing and beautiful.

One of the richest and darkest reds is seen in the Japanese herb shiso (also called perilla.) Its frilly, almost metallic-toned, nettlelike leaves are strikingly similar to purple basil. They are bland tasting but make a salad look very pretty. The green variety, which is used in Japanese cooking, has a much stronger taste. Sow shiso seed in spring in pots or cells, and don't plant out until all risk of frost has passed.

Enamel flour and bread bins were commonly used in kitchens until late last century and can still be found in flea markets and secondhand stores. Often brightly colored or printed with vintage text to describe the original contents, they make adaptable and unusual planters for everyday salad crops.

greencare

There are many varieties of red lettuce. A loose-headed type is ideal to grow in a small planter, because you can cut the stem above the soil and new shoots will be produced all around the cut. Allow these to grow, and keep cutting throughout the summer. Lettuces can be sown successfully from early spring until late summer to ensure a continuous supply; as long as the weather is not too cold, they will always grow. Continue to grow in a cloche or coldframe to extend the season.

Since they grow quickly, lettuces can be planted in smaller containers, but they need plenty of moisture and good drainage. Give them a very dilute liquid feed every week to add nourishment to the soil.

Above left: *Loose-headed red crinkly lettuce will thrive and grow in an enamel bin, but don't stand it in continual hot sun, or the tender leaves will wilt.*

Right: *Lovely vintage kitchen containers, often printed with descriptive text, are relatively easy to come by. The unusual dark lustrous leaves of the shiso mean it is grown more for decoration than for the kitchen, but it is a reliable plant.*

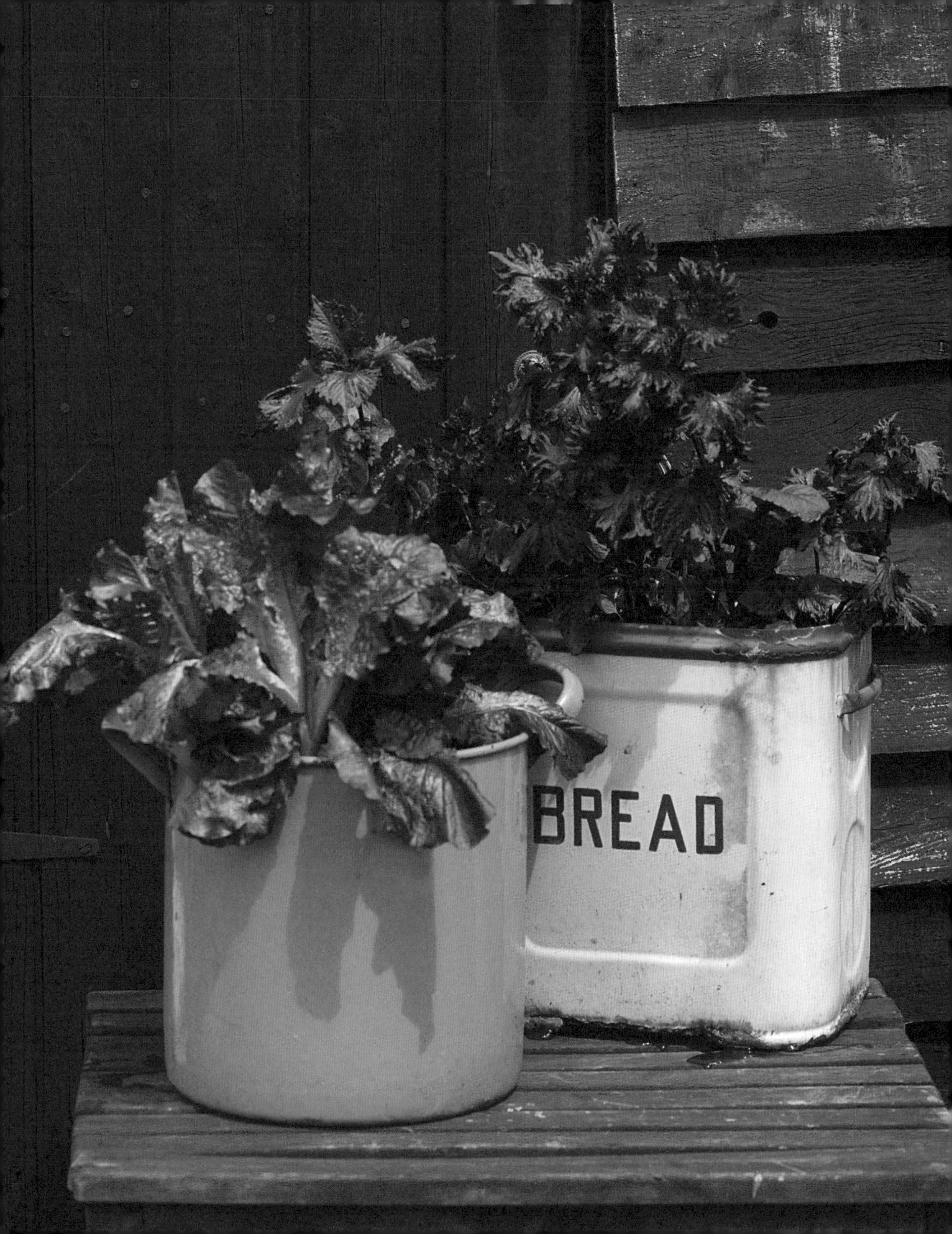
BREAD

Salad bowls

you will need

two plastic dishwashing bowls 18 inches (45cm) in diameter and 7 inches (17cm) deep

organic peat-free multipurpose potting mix

pea gravel

cordless electric drill

selection of salad plants, including red and green loose-leaf lettuces, giant red mustard, and mizuna

purple slate as a mulch

green*care*

Salads need less soil than some crops, but they must have enough to stop them from getting dry or parched. A dry, thirsty lettuce will become bitter and run to seed more quickly. Be careful also not to leave salad crops in the hot sun— this will make them wilt, and it may be difficult to revive or refresh the affected leaves.

Right: *Two large plastic dishwashing bowls planted with a variety of salad leaves will be enough to provide you with a salad every day. To ensure a continuous supply, plant a couple more bowls a month or so later.*

Even if you don't grow any other vegetables in your container garden, do try a few salad varieties. It is possible to grow enough salads in various containers to keep you going all through the summer and well into the fall. The best ones to grow are the cut-and-come-again types—harvest a few leaves to eat and simply wait for the plant to grow more. You can keep a crop going for some time, but it would be wise to sow seed or plant seedlings a month later in another pot to make sure of a continuing supply.

There are plenty of seeds available for cut-and-come-again salad crops, but a quicker alternative is to buy some young seedlings and plant them, correctly spaced, directly into the potting mix. The loose-leaved varieties are most suitable. Green and red lettuces look good together both in the pot and on the plate. For added taste and variety, grow a selection of oriental salad greens. These generally do better when planted later in the summer. Giant red mustard, with its peppery hot flavor and pretty red-veined leaves, is a "must" in a container garden. Combine it with the milder cut-leaved mizuna.

Large plastic dishwashing bowls are ideal salad planters. They are inexpensive and come in a variety of colors and sizes. You will need to make some drainage holes in the base, but there will be no evaporation through the sides.

1 Turn the plastic bowl upside down, making sure that it is on a firm surface. Drill a number of holes in the base about 4 inches (10cm) apart from one another.

2 Fill the base of the bowl with a layer of pea gravel to an even depth of about 1 inch (2.5cm).

3 Next, add the potting mix, filling the bowl to just below the rim. Break up any large clumps of mixture.

4 Select your lettuce seedlings, planting the red and green ones alternately in a circle. They should be about 4 inches (10cm) apart. This is not a critical measurement, but they should not be planted too close to one another, or overcrowding will check their growth.

5 Carefully arrange the slate pieces around the young plants, taking care not to damage the delicate leaves. The potting mix needs to be well covered; this will stop any weed growth and keep the mixture moist and cool.

6 Water well all around each plant, and keep them well watered throughout the season as they grow.

Left: *The red-veined leaves of giant red mustard and finely cut mizuna leaves provide decorative and tasty additions to a mixed-leaf salad.*

Rainbow radishes

Radishes are usually grown as a quick-maturing crop between slower-growing vegetables. Crisp and peppery, with beautiful white flesh, they are best eaten young, when the flesh is tender and the white, pink, yellow, purple, or red skins have not yet become tough. Radishes will grow easily in containers, but make sure that they are shaded from hot sun. A dry soil and too much heat will cause the radishes to bolt, making them woody and inedible.

After sowing from seed, you will have to thin out the seedlings to encourage the radishes to form—they are swollen stems and need room to grow. A gap of an inch or less between plants is fine, as long as you harvest the radishes when small. It takes only a month for radishes to mature, which makes them an ideal crop to interest young children.

In this case, the radish seeds have been planted in square plastic serving bowls, which are inexpensive and available in a good choice of vibrant colors. Before planting, simply pierce the base of each bowl a few times to make some drainage holes.

you will need

plastic serving bowls about 12 inches (30cm) square and 7 inches (18cm) deep

electric drill or craft knife

pea gravel

a rich loam-based potting mix

radish seeds, 'French Breakfast'

Left: *Pull the radishes gently out of the potting mix as they mature, leaving others more space to grow. Wash, trim off the roots and leaves, and serve immediately with a little flaked coarse salt.*

greencare

Experiment with some of the many different varieties of radish available, by sowing a new variety every two weeks during the summer and autumn months.

1 Make a few holes in the base of each bowl. This is easily done with a drill or, if the plastic is not too rigid, by cutting with a craft knife. Add a 1-inch (2.5cm) layer of gravel to promote good drainage.

2 Fill the bowls almost to their rims with potting mix, breaking up any large lumps and then firming it down in the bowl with the palm of your hand so that it is not too loose.

3 Sprinkle the seed thinly on the surface of the potting mix. Then tuck the packet down between the soil and the side of the bowl, so that you can identify the variety of radish when it is ready to pick.

4 Take a small amount of potting mix in your hands, and gently rub them together over each bowl, allowing a fine layer to cover the seeds.

Right: *If the seeds are sown regularly, at intervals of two weeks or so, you will be able to harvest enough radishes to keep you supplied for many months of the year.*

5 Water the surface of the soil gently with a fine spray, so as not to dislodge the seeds or wash them to one side of the bowl. Place the bowls away from full sun and keep the surface of the soil moist.

6 When the seedlings have reached a height of about 3 inches (7.5cm), thin them out, allowing a space of up to 1 inch (1-2cm) between each radish.

Colorful chard

One of the prettiest leaf vegetables to grow, chard comes in many colors. The vibrant pink stems of ruby chard are almost luminous when the sun shines through them. 'Bright Lights', the variety used here, is a mixture of pink, red, yellow, orange, and white stems; these are less vigorous than other types (which can grow to be enormous) but are suitable for container cultivation.

Swiss chard (sometimes also known as seakale beet) is the parent of all the colored chards. It has glossy, dark green, rather puckered

Right: *The dazzling color of the chard stems is echoed in the leaf veins.*

Below: *The simplicity of this white fiber clay planter, shaped to fit on a windowsill, is the perfect way of showing off the dazzling leaves of 'Bright Lights' chard. Fiber clay is lighter than fired clay but more substantial than fiberglass.*

leaves and brilliant-white, broad-ribbed stems. The stems last a long time and will even carry on producing substantial leaves through a mild winter. Just a few chard plants will be enough to feed a family. They have deep roots and won't be happy in a shallow planter.

Chard is tolerant of most soils, but you will get a better crop from a rich potting mix including well-rotted manure. Luckily, the plant is relatively free of pests, although in a wet season slugs and snails will enjoy taking a bite out of the young leaves. You could spread a light-colored mulch of crushed seashells over the surface of the soil after planting. This prevents too rapid evaporation, reducing the need for watering.

Chard leaves should be cooked separately from the stems, because they have different densities. The leaves can be used like spinach or sautéed and combined with olive oil, chili, garlic, and perhaps a little tomato for a wonderful pasta sauce. Braised stems can be served with a cheese sauce. Steamed stems cut into sections, cooled, and served as a salad with olive oil and lemon juice, are a simple summer favorite.

green*care*

You can sow chard seed in late spring for harvesting during summer and fall, and again later in the season, producing a continuous supply of this versatile vegetable through the winter until the early summer of the next year. Either sow *in situ* or sow in cells and transplant to the final position when the seedlings are about 2 inches (5cm) tall. They should be spaced at least 4 inches (10cm) apart—more if you want sturdier and stronger plants.

Spinach in a supermarket basket

Real spinach (as opposed to spinach beet, which is a close relative of chard) is a tender and sweet-tasting leaf crop. Spinach is best grown in the spring and fall; in hot weather it tends to run to seed. Don't grow spinach in hot sun—a lightly shaded position will produce a healthy crop. Watering is very important; spinach does best in a loam-based, moisture-retentive potting mix.

In this case, spinach plants have been planted in an old supermarket basket. The regular grid pattern of the wire creates a pleasing decorative effect. The basket has been lined with a hanging-basket liner that incorporates a perforated plastic film on the inside to prevent too much evaporation through the wire frame. (Alternatively, add a perforated sheet of plastic inside an ordinary liner.)

Use a rich potting mix, adding some well-rotted manure. A mulch of pea gravel spread around the base of the plants will help to retain moisture and keep the soil cool.

greencare

You can sow spinach direct into your container. Alternatively, sow a few seeds in a small pot of seed-starting mix, and transfer the clump of seedlings into their final growing position when they are about 1 inch (2-3cm) tall. Thin the clump, leaving one or two plants together (the thinnings make a sweet addition to a green salad).

Carrots and beets in jazzy tubs

you will need

gimlet (for boring holes)

dibble

plastic or rubber flexible buckets, about 12 inches (30cm) high and 14 inches (35cm) in diameter

soilless multipurpose potting mix with some added loam-based mix

beet (beetroot) and carrot seeds

fine sand

green*care*

Beets, with their sweet, earthy flavor, are an undemanding crop that can tolerate most conditions, although it needs regular watering in dry spells. If seed is sown too early, the plant may bolt; you could try a variety called 'Bolthardy', which is bolt resistant.

Carrots and beets are both well suited to container growing, and their vibrant colors make a brilliant contrast when grown in colorful plastic or rubber buckets.

Eaten straight after harvest, either raw or lightly steamed, homegrown carrots retain all the sweetness that is so often missing from store-bought ones. You can sow the fast-maturing varieties every few weeks to ensure a supply throughout summer and fall. Sow the first crop in early spring, first mixing the tiny seeds with a little fine sand to make sure they are sown more thinly. As they grow, the seedlings should be thinned to avoid crowding and to allow the remaining plants to grow bigger. Water well and often, applying a liquid seaweed feed every two weeks to improve the yield.

There are many varieties of beet available, some with exotic names. 'Bull's Blood' is an old reliable type, while the stunning 'Chioggia' is fuchsia pink with paler stripes and sweet tender flesh. There are cylindrical types, which are easy to slice, and beautiful golden globes, such as 'Burpees Golden', which look wonderful on the plate. Sow from spring to midsummer for a continuous supply.

The leaves and stalks of beet are also good to eat. Eat them young in salads, or chop and cook the older ones as you might spinach or chard—sauté them in a pan with olive oil and some chopped chili and garlic, and serve with fresh pasta.

Above left: *It's easy to see when beets are ready for cropping, and all parts of the plant are edible, so nothing is wasted.*

Right: *These colorful rubber containers will hold a good deal of potting mix, which makes them ideal for growing carrots and beets. Sow seed in the tubs where they are intended to grow, as they will be too heavy to move later if you change your mind.*

1 Turn each bucket upside down and make a few drainage holes in the base with the gimlet.

2 Fill the bucket to the rim with the mixed planting medium. The mixture will settle down somewhat when watered.

3 Use the dibble to make small holes in the mixture, 2 inches (5cm) apart. Sow a few beet seeds into each hole.

4 Mix the carrot seeds with a little fine sand, and sprinkle them thinly onto the surface of the mixture. Push the seed packets into the edge of the mixture to identify the beets (beetroot) and carrots.

5 Water well, using a watering can fitted with a fine rose, and wait for the first seedlings to emerge.

6 When the beet (beetroot) seedlings emerge and develop their first true leaves, thin them again to give the plants more growing space. Leave the carrots to grow for a little longer before you start to thin them.

Left: *The combination of the bright pink and deep purple of the beets looks stunning against the pink container, especially when the evening sun shines through the stalks of the young plants.*

summer favorites

New potatoes, tomatoes, eggplants (aubergine), zucchini (courgette), bell peppers, and chilies are ready to harvest in summer.

Don't compare your crop of bell peppers to those you find in stores. Their uniform shape and unblemished appearance reflect the unnatural conditions in which they are grown. Your organic, slightly small, possibly misshapen peppers are probably more "natural" than any peppers you can buy. Peppers are adaptable plants and thrive in the most unusual containers, as long as they have enough rich soil and plenty of moisture. Place peppers indoors on a sunny windowsill if the weather is unfriendly, but remember to keep them well watered and fed.

It is possible to grow a whole year's supply of fiery chilies from just a few plants. They can be eaten fresh or strung together and hung up to dry above the stove or in the sun for later use. Since a long season from sowing to harvest is necessary to allow them to ripen fully, I recommend buying young plants of named varieties—these will have been started off by a professional grower early in the year in optimum conditions. Like bell peppers, chilies need heat and humidity, and benefit from a fine-mist water spray morning and evening. Bring them inside when the nights get cold.

Zucchini (courgette) are also easy to grow. They need good rich soil and plenty of space. There are several varieties of zucchini (courgette), from dark green to pale green striped; there are even golden zucchini (courgette) (slightly less productive than other varieties) and curious round ones. Pick them young to encourage the plant to continue producing. The large golden flowers are a delicacy in Italian cooking, but pick them only after the tiny zucchini (courgette) is visible behind (which means that the flower has been pollinated). One option at the end of the season is to allow some zucchini (courgette) to remain longer on the plant and grow into marrow squash. These are not as sweet as young zucchini (courgette) but are very good when stuffed, and economical too—one stuffed marrow squash will feed four people.

Potatoes in woven sacks

you will need

polypropylene potato-growing sack or similar

seed potatoes: an early variety, such as 'Red Norland' or 'Charlotte', and a keeping variety, such as 'Yellow Fin' or 'Pink Fir Apple'

good potting mix combined with some well-rotted manure

egg boxes or tray

Is there any reason for growing your own when potatoes are so readily available and cheap? It's not that you can't buy organic potatoes, or that you can't find the variety you want. It is for the pure pleasure of the experience—the satisfying process that begins with the early sprouting of the seed potatoes and moves on to the planting in rich potting mix, the earthing up, the anticipation of the harvest. If you cook them within minutes of the harvest, you will be in no doubt about the value of the exercise. Store-bought potatoes taste utterly different from those that are sweet, fresh, organic, and homegrown.

Not much space is required to grow a few potatoes, since they are easy to cultivate in all sorts of containers. Builders' sand bags or plastic garbage cans are commonly used. This project uses purpose-made woven polypropylene sacks; the loose style of weave provides good drainage. Potatoes are packed for transport in sacks made from a lighter version of the same material—you could ask a local greengrocer to save some for you.

Right: *Potatoes grow happily in big plastic woven sacks and produce huge numbers of leaves. These help to keep the soil weed free and moist, encouraging a bigger crop of potatoes.*

Potato varieties

Potatoes are from the same family as eggplants (aubergine) and tomatoes. There are hundreds of varieties of potato to choose from, so consult a seed catalog to see which tastes and textures appeal to you. A potato variety will often be described as "floury" or "waxy"; I think the waxy types are more adaptable in the kitchen.

Potatoes can be grouped roughly into two types: early-maturing and keeping potatoes. The former are ready to be harvested about 7 to 9 weeks after planting, while the latter (also called late-season) are harvested within about 10 to 12 weeks.

A special small category, called fingerlings because of their small, elongated tubers, requires three months of frost-free weather to reach full maturity. That said, you can harvest most potatoes more or less when you want, once the plants flower. Leaving the plants in the soil a little longer will give you bigger potatoes.

As a general rule, potatoes grow best in a cool climate, where the summer temperature rarely exceeds 90° F (32 °C). Your local garden center or nursery can advise you on the cultivars best suited to your climate and growing conditions.

1 Lay the seed potatoes in the egg boxes or tray and place in a cool, light, and frost-free place to encourage the tubers to sprout, or "chit." They are ready when the sprouts are short and dark green or purple, which usually takes about four to six weeks. A shortage of light will result in weak, etiolated shoots, which will affect the growth of the potato plant. Chitting gives the potatoes a head start and helps to produce a larger harvest.

2 Roll down the sides of the sack so that you can easily reach the base. Add potting mix to a depth of 4–6 inches (10-15cm) above the base. Choose five or six seed potatoes, and from each rub off all but two or three sprouts at one end. Place the potatoes, sprouted end up, on top of the soil.

3 Bury the seed potatoes 6 inches (15cm) deep under more potting mix, and wait for the sprouts to grow and break through the earth. This could take a couple of weeks—or longer, if the weather is very cold.

4 As the plant grows and the leaves emerge, cover them again; this is called earthing up. It is partly a precaution against frost for early-planted varieties and partly to encourage lots of tubers to form along the stems, giving a bigger crop. As the crop grows, continue earthing up, unrolling the sack as necessary. In dry periods, make sure you give the plants plenty of water.

5 When the potato plants begin to flower, you know it won't be long before you can harvest them.

5 Dig out a sample potato to see what size they have reached, and leave the rest a little longer if you want larger tubers. Either tip out the soil and potato crop from the sack or dig out a few as required. Keeping potatoes should be harvested when the foliage dies back.

chitting

To chit seed potatoes is to leave them in a cool, light place to encourage sprouting. Chitted potatoes produce a quicker, heavier crop—in effect, it means that the early part of the growing process is speeded up. Put the seed potatoes in egg boxes and place them in a porch or on a cool windowsill. During the next few weeks, stout green or purple shoots will appear at the top of each potato. Only three or four of these are needed, so you can rub off the rest at the time of planting.

green*care*

Potatoes need regular watering to produce healthy tubers. Plant them in a rich soil, mixing some well-rotted manure into the potting mix.

Potato blight, which is a serious disease, can be largely avoided by growing early varieties. If the leaves look burned, to avoid damage to the tubers, cut off the top growth. The tubers can then be dug up normally.

To preserve optimum flavor, harvest your early potatoes minutes before you are ready to cook them.

Vining tomatoes in a blue plastic bucket

you will need

large plastic bucket

gravel or crocks for drainage

three varieties of young tomato plants

organic soilless multipurpose potting mix with some added loam-based mix

fresh comfrey leaves (optional)

tray of French marigolds

Normally classified as indeterminate tomatoes, vining tomatoes need a deep, spacious container, especially if you grow more than one variety together. A rich, well-balanced potting mix is essential. To ensure a healthy crop, when the fruits mature, give the plants a weekly liquid feed; an organic tomato or seaweed feed is ideal.

Pinch out the side shoots as they appear between the stem and the leaves. Allow about five sets of flowers, then pinch out the growing tip, known as the leader. This stops the tomato plant from growing taller and producing more flowers and ensures that all the plant's energy goes into ripening the fruit.

Tomatoes are easy to grow from seed and make a great plant for novice gardeners to try. Start them off early in the year in small pots sited in a warm place to encourage growth, then transfer the seedlings into larger pots, so that the young plants can become sturdy enough to grow outside when all risk of frost has passed. It is important to get them going early, so that you have a long cropping season during the warmest weather.

The large blue plastic bucket with rope handles used here is just the sort of container that's discarded every day in dumpsters and town dumps. Buckets of this kind are common on building sites, so keep your eyes open—you never know where you might come across containers that have great potential for your garden.

Tomato varieties

Homegrown tomato plants seem to produce tomatoes that are much more delicious than those on offer in supermarkets—and the fruit has that wonderful just-picked smell. Another great advantage of growing your own is that it allows you to try out different varieties that will never be available commercially. Reading the tomato section of a seed catalog is a mouthwatering experience, and it can be hard to narrow down the list of varieties that interest you.

The appearance of the mature fruit may be a factor in your choice. Tomato colors range from deep orange and vermilion to clear yellow. Some varieties have stripes or blushes. There is also a pretty pink variety as well as a number of curious black versions.

Even more fascinating is the choice of shapes—from the huge beefsteak to the tiny cherry tomato. The elegant plum tomato, commonly used in Italian sauces, is particularly rewarding to grow.

Right: *Growing French marigolds underneath tomatoes is a well-tried and effective form of companion planting. The pungent smell of the marigolds repels aphids, so organic gardeners plant them all around the vegetable beds. The vibrant orange of the flower heads makes a striking partner for the tomato trusses above.*

blight

One of the worst enemies of the tomato is blight, which is a nasty disease that the tomato shares with its close relative the potato.

Blight is particularly prevalent during damp, humid summers. Plants that have been infected look as if their leaves have been burned, while the fruits have blackened areas and become inedible.

The best way to avoid blight is to grow your plants under cover. This may not always be possible, but growing tomatoes in a container, away from other blight-affected plants, will help. In order to prevent an attack, you can try spraying plants with a copper fungicide (Bordeaux mixture).

Right: *Planting vividly colored marigolds at the base of the tomato plants makes an exotic display, as well as protecting the tomatoes against an unwelcome aphid attack.*

1 Make a series of drainage holes in the base of the bucket. Add a layer of gravel or crocks and fill the bucket with the potting mixture.

2 Add the young tomato plants, spacing them evenly in the mixture. Take care not to damage their roots when removing them from pots.

3 Leave the plant labels beside the plants so that you can identify your crop later on and compare its taste with other tomatoes.

4 Plant the marigolds all around the tomatoes. As the tomatoes grow tall, the marigolds will bush out and remain low, covering the soil. Vining tomatoes need staking. Use sturdy canes pushed into the soil beside the plant. Tie the stems to the canes regularly as they grow.

green*care*

As the tomato fruits begin to ripen, remove a few of the lower leaves to allow the sun to reach the fruits, which will encourage the ripening process.

Bush tomatoes in a wire basket

Tomatoes, although strictly fruit, are probably the most popular "vegetables" to grow; and, as is evident from any seed catalog, there is an unbelievable variety to choose from.

Tomato plants fall into two basic categories, describing their growth habits: indeterminate, or vining, and determinate, which have a more compact, bushy form. The determinate is probably the easier to grow, because it doesn't need "pinching out," or pruning. The vining type, which produces a heavier crop, will grow until you pinch out the top shoot; it needs to be tied in to a sturdy stake and should have all the side shoots removed (see page 82). A special hybrid, called 'Tumblers', is small and bushy and ideal for growing in hanging baskets (see page 88).

Window boxes or troughs are good for the more compact varieties of tomato; large pots or buckets are suitable for the indeterminate, vining types; and the determinate kind will thrive in a variety of containers. Wire ones must be lined to conserve moisture. Here, the lining consists of a roll of pressed felt specially designed for hanging baskets; it has a perforated plastic film on the inside and can be cut to fit any container with an unusual shape. Alternatively, an ordinary liner can be given an extra layer of plastic.

green*care*

Tomatoes need a good rich potting mix. Use a multipurpose potting mixture and add a good loam-based growing medium. This aids moisture retention, which is one of the most important requirements when growing any crop in a container of any kind.

The plants also need a good soaking of water every day during hot weather and will benefit from a special tomato feed once a week when the fruit has set.

Most importantly, they need sun; in a cool summer, it will be difficult to get all your tomatoes to ripen, but the unripe green ones make the most wonderful chutney.

Above left: *Plenty of plump tomatoes ripening on the vine makes one of the most magical sights of summer.*

Right: *The tomato has been planted in an old potato harvesting basket and has been underplanted with some bush basil plants. This is a small-leaved compact basil, often referred to as Greek basil, a perfect and traditional herb to accompany lovely ripe tomatoes.*

'Tumbling Tom' in a hanging basket

you will need

wire hanging basket of your choice (plus chain and hanging bracket, if necessary)

sphagnum moss for lining basket

rich loam-based compost

water-retaining granules

tomato plant ('Tumbling Tom', red)

spare bucket (to rest basket in)

'Tumbling Tom' is a small, prolific tomato plant that's ideal when you don't have much space. It will grow happily in any kind of container, as long as it is planted in a rich mixture and fed from the time when the fruits begin to mature. Two or three plants in a window box will give you a heavy crop of sweet cherry tomatoes all summer.

Cherry tomatoes are perhaps the easiest tomatoes to grow. Being compact bushes, they don't need any pinching out. For tomatoes to thrive outside, they need to be in a warm place—against a sunny wall, for example, where the stone or brick will retain warmth after the sun has disappeared at the end of the day.

If you grow this variety of tomatoes in a hanging basket, you must be vigilant about watering and feeding. It may help to add a small quantity of water-retaining granules to the loam-based mixture to ensure that the soil stays moist. In hot weather, water twice a day. Purpose-made hanging baskets of various kinds are available at garden centers. The one used in this project is a vintage wire basket that is extremely decorative. Hang the basket from an old cast-iron bracket, if you can find one.

green*care*

Sow the seed from late winter until early spring in individual pots, and cover with potting mix to exclude the light. If you keep the pots in a warm place, the seeds should germinate in a week. Choose the sturdiest seedling in each pot and discard the rest. Allow it to grow, putting the young plant outside to harden off only when the weather warms up. If there is danger of frost at night, bring the plants indoors.

Tomatoes need at least three months of warm weather (ideally, between 70° (21 °C) and 80° F (26°C), during the day and about 10° lower at night) to produce sweet fruit. Rain, wind, and variations in temperature can make the leaves look bedraggled. Very hot temperatures (above 100° F (37°C)) make pollen infertile and suspend the growth of the fruit. By choosing an early-season and a mid-season cultivar, you can enjoy homegrown tomatoes for many months.

Left: *Little cherry tomatoes are a treat to eat straight from the plant.*

Right: *Even one small plant grown in a hanging basket or window box will provide you with a respectable crop of tomatoes during the summer.*

1 Line the hanging basket with the moss, creating a thick, even layer around the sides and base and filling any gaps.

2 Rest the basket in the spare bucket. Put some potting mix in the base, and add a small amount of water-retaining granules (follow the manufacturer's instructions).

3 Fill the basket to the top with potting mix, and make a hole in the center large enough for the tomato plant.

4 Place the tomato plant in the hole and secure it by firming it in with your fingers. Water well.

comfrey liquid

Fruit-producing plants such as tomatoes need a high-potash feed, that is one containing potassium, to encourage the plant's development and stamina. Specially formulated organic feeds are available, but you may like to make your own comfrey liquid if you have access to comfrey plants (see page 16). Comfrey is rich in minerals, and particularly high in potash.

Fill a suitable container, such as a rubber bucket, as here, or an old enamel bucket, with freshly cut comfrey leaves—pack them in tightly, cover with water, and weight them down with a large stone, or something similar. Cover the bucket and set it in a shady out-of-the-way place. The reason for this is that the smell is truly awful.

After three weeks or so, decant the liquid into plastic milk bottles. (This is an excellent way to reuse plastic bottles.) Be careful to label them clearly.

When you come to use the liquid, first dilute it 1:5 in water. Once the flowers have set on your tomato plants, give them a good feed of this liquid every week.

Freshly cut comfrey leaves make a great mulch around the base of tomato plants. You could even chop some up and mix into the planting mixture.

Eggplant (aubergine) in a rubber bucket

you will need

recycled rubber bucket

craft knife

multipurpose potting mix combined with loam-based mix and a couple of handfuls of well-rotted manure

young eggplant (aubergine)

6 marigold plants

Eggplants (aubergine) have amazing purple-black fruits with skin as shiny as a mirror. The downy stem and leaves are unlike those of any other plant, although they are related to the tomato family. The plants may appear to be soft all over but take care when handling because there are some sharp spines, especially on the calyx at the top of the ripening fruit.

Eggplants (aubergine) need a long spell of hot weather, and so are ideally suited to most parts of the United States. Although some varieties are tolerant of cooler climes, they do not appreciate great fluctuations in temperature. Cold nights are not helpful. If you grow eggplants (aubergine) in a container, place it against a warm wall or similar outside space. The heat absorbed from the sun during the day will radiate at night, helping to create good eggplant (aubergine)-growing conditions. The rubber bucket used in this project is made from recycled tires.

If the plant grows well, it may need staking to prevent wind damage, especially if it is bearing several heavy fruits. If the weather is too cool or there are too few pollinating insects, the flowers may not set fruit. One way to alleviate this problem is to spread the pollen from flower to flower with a small paintbrush. You can tell if pollination has been successful because a tiny swelling, which will eventually become the fruit, appears in the center of the flower.

green*care*

Underplanting with French marigolds helps to deter aphids and keeps the eggplants (aubergine) healthy. The marigold flowers will not discourage slugs.

I recommend buying in young plants—a nursery will have started them early in the year in large, heated greenhouses. It is becoming much easier nowadays to buy healthy organic plants for very reasonable prices.

Above left: *Exquisite lilac-colored flowers, which are particularly inviting to bees, precede the distinctive, deep purple fruit.*

Right: *You can't help feeling proud when your eggplant (aubergine) produces a beautiful shiny fruit. A rubber bucket makes a deep, roomy container for this lovely vegetable.*

Right: *A bright green can, which once contained olives, makes a useful and unusual container for this beautiful eggplant (aubergine). I didn't even have to cut away the top, since it had a large circular lid that, when removed, left a perfect planting hole.*

1 Cut some drainage holes in the bucket's base. It is simplest to cut a triangle, with each side ½ inch (1cm) long. Fill the tub with the potting mix to within 1 inch of the rim.

2 Make a hole in the center. Remove the eggplant (aubergine) from its pot and place in the hole. Firm the mixture gently around the plant.

3 Plant the marigolds equidistantly around the eggplant (aubergine). To make the marigolds bushier, pinch out the central growing tip—this will encourage the plant to produce more side shoots, with more potential for flowers. As the eggplant (aubergine) grows, add a stake and tie the stem to it with garden twine at regular intervals.

Right: *The skin of a ripe eggplant (aubergine) is tight and shiny— like a curved mirror, it reflects the dazzling orange flowers of the companion marigolds.*

Chilies in olive-oil cans

you will need

olive-oil can

can opener

hammer

thick, long nail

terracotta crocks or gravel

multipurpose potting mix combined with loam-based mixture

young chili plant

Chilies can be really prolific plants, and it is quite possible to grow enough to last you all year. You can use some of the fresh chilies as they ripen and then dry the remainder for storage and use them throughout the winter—simply lay them out on a tray covered with a clean cloth and leave in a warm place to desiccate slowly. Alternatively, thread them together in bunches on a long piece of string and hang them in a warm, dry place.

Chilies will thrive in warm weather outside, but they generally prefer quite humid conditions. When growing chilies inside on a sunny windowsill, keep them damp by watering with a fine spray. If the plants or soil dries out, they are likely to drop their buds.

These colorful and exotic plants deserve to be grown in equally decorative containers. Empty olive-oil cans printed with interesting text and images are ideal. You can ask a local restaurant or health-food store to save them for you; in this way, you could build up quite a collection—enough to contain a whole family of assorted chili plants. These metal planters will last for a few years outside before rust penetrates and they begin to fall apart.

Right and below: *Chilies grow quite happily in containers. A number of different cans planted with a few varieties of chili makes a decorative display—and these plants are a useful crop to grow. Dry your chilies at the end of the season, when they are ripe, and you will have enough to last until the next season.*

green*care*

The seeds germinate rapidly and can be sown in early spring inside. Sow in small pots and prick out the seedlings, leaving the strongest one in each pot. When the weather warms up, transfer the young plant into your chosen container. Use a rich potting mix, and as the small chilies start to form, feed with a high-potash liquid feed.

If you don't have time to grow chilies from seed, it is easy to find named varieties in a good plant nursery. They will be raised and sold alongside their close relatives: peppers, tomatoes, and eggplants (aubergine).

Only you will know how hot you like your chili—some are sweet and only slightly fiery, while others are unbearably hot. Seed catalogs indicate the intensity of different varieties, and young commercially grown plants will often be labeled with the level of heat.

I recommend growing a variety of types of chili, from mild to hot, and a variety of shapes and colors, from red, green, and yellow to the more unusual black.

1 Remove the top of the can with a can opener. Smooth any sharp burrs on the metal rim by banging a hammer around the rim. Make four holes in the base with the nail.

2 Turn the can the right way up and put a 2-inch (5cm) layer of broken crocks or gravel in the base. This will encourage good drainage.

3 Fill the can to the rim with the potting mixture. Shake it down and then add some more. Break up any large lumps of mixture.

4 Make a hollow in the mixture. Tap the chili plant out of its pot, and plant in the can. Firm in the plant with your hands. Water well.

green*care*

As the plant grows, you may need to use a small stake to support it, because the chilies can become quite heavy. If you keep picking the chilies, the plant will continue to produce more. To prolong the harvest, bring the plant inside at the end of the summer.

Right: *These fiery red chilies have been planted in a small galvanized bucket. They are ripe when the skin begins to wrinkle slightly; pick and use immediately, or dry and store for future use.*

Red and yellow peppers in a plastic basket

I love the intense colors and intricate patterns of woven plastic baskets. They are also very light and extremely durable—both desirable qualities in a planter—and many have handles, which makes them easy to move.

This basket is lined with a green refuse sack; the top of the sack is rolled down and hidden behind the rim of the basket after the potting mix has been added. Don't forget to cut a few slits in the base of the

sack and add a few pieces of broken polystyrene to promote good drainage.

It is advisable to buy bell peppers as young plants. To raise them from seed, you need to sow very early in order for the plants to be advanced enough to grow and ripen during the summer. The hotter the temperature, the quicker the peppers will mature, of course. If your area has cool summers, you could grow this plant on a sunny windowsill to make the most of the heat of the sun.

Bell peppers come in many shapes, colors, and sizes. Don't be surprised if yours don't look like the ones you can buy. Store-bought peppers are always grown in controlled conditions with little fluctuation in temperature, and any misshapen fruit will be discarded, rather than being offered for sale.

greencare

As with most container-grown produce, bell peppers should be watered often and, especially during hot weather, in the evening. This deters the roots from rising to find the water and getting scorched near the heated surface of the soil. Give a weekly liquid feed once the fruit has matured.

Above left: *Pick the fruit as the color deepens, because this encourages more bell peppers to form.*

Right: *A couple of bell pepper plants will thrive in this colorful woven bicycle basket. They need a warm summer and moist conditions.*

Zucchini (courgette) in a galvanized washtub

you will need

- old galvanized metal washtub
- hammer and large nail
- bucket of gravel (for drainage)
- multipurpose soilless potting mix and rich loam-based mix
- well-rotted horse manure
- bucket of comfrey leaves (if available)
- organic fish fertilizer
- two or three zucchini (courgette) plants

Vintage galvanized domestic and agricultural items, such as washtubs, buckets, trays, watering cans, and grain bins, are becoming sought after as practical planters. There is something about the gentle gray of the metal and the pleasing shapes and proportions of these objects that makes them ideal containers for plants. They are lightweight and good at retaining moisture—none evaporates through the nonporous sides.

Zucchini (courgette) are large, thirsty, hungry plants that need a large container—a washtub is an ideal size. They are easy to grow, and given the right conditions, they romp away, creating a virtual jungle of stems and leaves. Zucchini (courgette) become slightly bitter if they are not eaten immediately after harvest. Pick them often—if they are allowed to grow too large, the plant will be less productive.

If you don't have enough space to start them inside, you can sow zucchini (courgette) directly into their permanent position; but wait until late spring. You could make a mini cloche from a plastic water bottle for added protection. Alternatively, buy young plants from a garden center or nursery or plant sale; many varieties are available, and since you will need only two or three plants, this is a sensible option.

Above: *The beautiful golden flowers of the zucchini (courgette) are edible and delicious. Pick them after the zucchini (courgette) has formed. The flowers are usually dipped in a light batter and fried in olive oil. In Italy they are a delicacy and are often served stuffed with anchovies or mozzarella cheese.*

Right: *Although a plant carries both male and female flowers, the zucchini (courgette) fruit forms behind female flowers only.*

Zucchini (courgette) varieties
There are many varieties of zucchini (courgette), ranging in color and shape from golden yellow (a less vigorous plant) to dark, shiny green, striped or pale-skinned, and from long to globe-shaped. Grow two or three varieties and see which you like. Start them off in spring under cover or inside. Push the flat seeds down into a seed-starting mix in individual pots—you can sow two seeds together and remove the weaker seedling as they grow.

1 With the washtub the right way up and resting on soft ground, make a number of drainage holes evenly in the base with the hammer and nail.

2 Empty the bucket of gravel into the tub and spread it evenly across the base. This layer will encourage good drainage.

3 Mix together the two potting mixes in equal quantities and empty into the washtub, covering the gravel to a depth of about 4 inches (10cm).

4 Collect the comfrey leaves and chop them roughly. Cover the soil with a thick layer of chopped leaves. This will provide a rich nutritional boost to the growing plants as the roots reach down.

green*care*

If you can't find fresh comfrey leaves to add to the potting mix, you could feed the zucchini (courgette) weekly with a diluted tomato or seaweed extract instead.

5 Fill up the tub to the rim with the combined potting mixes, this time adding one part well-rotted manure to four parts mixture.

6 Sprinkle the organic fish fertilizer on the surface of the soil. (First read the manufacturer's instructions, and don't use too much.) Lightly dig it in.

7 Plant the young zucchini (courgette) equidistantly from one another, and firm down the soil around them. Water well. Keep the mixture moist, and never let the soil dry out.

Right: *Zucchini (courgette) love warm weather and will soon produce a canopy of large leaves on sturdy hollow stems. The leaves shelter the growing vegetables from the direct heat of the sun and reduce evaporation of moisture from the soil.*

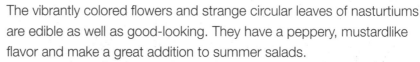

Nasturtiums in a metal bucket

*Right: **The rich vermilion of the bush variety of nasturtiums contrasts well with the chalky-blue painted galvanized mop bucket.***

*Below: **Use the open young flowers and small leaves as a colorful and peppery addition to a fresh summer salad.***

The vibrantly colored flowers and strange circular leaves of nasturtiums are edible as well as good-looking. They have a peppery, mustardlike flavor and make a great addition to summer salads.

There are many kinds of nasturtium, from compact, bushy plants to trailing or climbing versions, and they come in a rich variety of warm colors, from brilliant orange to rich mahogany. Sow the large seeds once, and you will have them forever, since they are prodigious self-sowers. They germinate quickly and can be sown until midsummer. Nasturtiums do best in poor soil; too much goodness encourages leaves at the expense of flowers.

Later in the season, you could grow a climbing nasturtium as a partner to a squash or a decorative gourd, allowing it to scramble up and over an arch, obelisk, or loose structure of canes or sticks. Don't plant too many nasturtiums, though, or their vigorous habit will overwhelm everything else.

green*care*

Nasturtiums are often grown as companion plants to tomatoes in order to attract aphids away from the main crop—aphids love nasturtium flowers even more than they love tomatoes. Fortunately, nasturtiums can grow even faster than aphids can eat them.

bean feast

Beans of all kinds are easy to grow. They suffer from few pests or diseases and tolerate most weather conditions. Another advantage is that they are highly nutritious—for some varieties it's best to eat the young green pods while for others you can leave them on the plant to ripen into beans to be dried for eating in winter.

Dwarf beans produce a smaller crop than their climbing relations. Sow at least twice during the summer to extend the cropping time, or pick the beans small and young. A plant must produce seed to propagate itself the following year, and regular picking encourages it to produce more.

Climbing, or pole, beans ripen from the base upward and crop for a few weeks from midsummer. Scarlet runner beans, in particular, with crimson flowers held high above the leaves as an invitation to bees, have often been grown for decorative purposes. Curiously, climbing

French green beans are pollinated by the wind and do not need insects. Beans for drying, such as crimson-speckled borlotti, can be left on the withering vine to ripen in late summer sun. Thoroughly dry them before storage to inhibit mold growth. To encourage drying, lay them on a tray indoors in a warm place for a few days.

The typical hot American summer is perfect for lima beans. Grow them directly from seed in a window box containing well-drained soil. Although pole lima beans require some form of support (hence the name), they have a higher yield than the bush variety. If your summers are cool, consider growing fava beans. You can plant seed directly in the soil when this has warmed slightly after the winter, or plant in biodegradable pots, which can be transferred into the soil.

Peas, too, require a cool climate. They also need a rich soil and should be well watered, but do not allow them to become over-wet. You may need to throw a net over the crop to protect it from birds.

Known collectively as legumes, beans and peas "fix" nitrogen in the soil by producing small nodules on the roots. At the end of the season, dig the roots into the compost and the nodules will feed the soil.

Dwarf green beans in a round basket

you will need

- **round cylindrical basket**
- **newspaper**
- **heavy-duty plastic garden refuse sack**
- **seed tray**
- **dwarf French green beans**
- **multipurpose potting mix combined 3:1 with well-rotted manure**
- **seed-starting mix**

French green beans will grow happily in a container, so long as the soil is rich and moist. They are prolific croppers and will provide a good harvest over four to six weeks. You can extend the cropping time by sowing more beans four weeks after the first batch; this is known as successional sowing. (You will need another container in which to plant this later crop.) As the second sowing will take place later in the year, when the air temperature is warmer, the beans can be sown directly into the container. The earlier crop should be sown individually in tall pots. When the plants start to flower, feed them weekly with a liquid manure or seaweed extract to ensure a long-lasting crop.

The deep, round basket used here is an ideal size in which to grow five dwarf green beans. It is lined with a garden-refuse sack, which stops water leaking from the basket—but don't forget to make some drainage holes in the base. Another advantage of using a basket is that it is lightweight—if you are growing crops on a roof terrace, you need to be careful about how much weight you are introducing.

Above: *Pick your beans just before you cook them for the best flavor.*

Right: *Don't be tempted to include too many plants in your container. Five make a really bushy group and will produce a good crop of lovely crisp, stringless green beans.*

1 Make some newspaper pots—ideally using a Paper Potter (see page 24). Fill each pot with seed-starting mix. Make a hole in the center with a stick, and push one bean into each hole. Cover the bean with mix.

2 Water the beans and put the pots in a plastic coldframe or bring them inside and place on a cool windowsill to allow the beans to germinate.

3 Push the plastic sack well into the basket and roll the edge around the top rim. Make a few slits in the bottom for drainage.

green*care*

Beans go well with summer savory, which resembles thyme. Coincidentally, summer savory can also be grown as a companion plant to beans, warding off aphids with its powerful aroma.

4 Place a few sheets of folded newspaper in the base to aid moisture retention. Add the well-mixed potting mix and manure, and fill the basket to the brim.

5 Remove the plastic sack from the rim of the basket, roll up and tuck inside beneath the soil.

Right: *When the flowers appear and the small beans start to form, you may need to give the plants some support. Tying them with string to a few bamboo canes will stop them from flopping over the edge of the basket.*

6 Plant the young beans in the basket, still inside their paper pots, when the first two real leaves have emerged. Space them widely—about five will be plenty in this size of basket.

7 Soak the beans with water poured from a watering can fitted with a fine rose, and place the basket in a warm, protected, sunny position.

Petit pois in galvanized buckets

you will need

biodegradable containers, such as paper pots (see page 24)

galvanized bucket

hammer

long nail

gaffers (electrical) tape

rich potting mix

well-rotted manure

old newspaper

comfrey leaves (if available)

pea sticks or bamboo canes

Right: Pick peas before they swell to fill the pod too tightly, or they will taste rather starchy.

If you buy peas, it is only worthwhile buying them frozen—freezing immediately after harvest preserves the natural sugars in the pea. For the same reason, homegrown peas eaten soon after harvest taste far better than so-called fresh peas in the pod from a supermarket.

Peas don't like hot weather, so are best grown in spring or fall. Sow the seed as soon as the soil is workable in spring or about eight weeks before the first frost is expected in the fall. A fairly deep container is needed, and you must water the plants regularly. They love a rich, moist soil. Newspaper at the base of the container will help to preserve moisture, and a layer of comfrey will add a wealth of nutrients, promoting healthy growth as the leaves rot down. Peas are leguminous plants, which add nitrogen to the soil through small nodules on their roots. At the end of the season, cut down the plants and leave the roots intact; this fertile soil can be planted with a late-cropping salad.

Peas also need a structure to grow up. As an alternative to bamboo canes, try the traditional British pea sticks (see page 116). These are simply twiggy branches (usually cut in winter when the stems have turned woody) from hazel, beech, hornbeam, or another suitable tree. Peas find them perfect for climbing.

1 Plant two or three pea seeds early in the year in a biodegradable pot. Keep them well watered and frost free, allowing the shoots to grow strongly.

2 Turn the galvanized bucket upside down and stick four short pieces of gaffers, or electrical, tape on to the base, spacing them evenly. Place the nail on each strip of tape in turn and bang firmly with the hammer to make a hole.

3 Turn the bucket right side up. Place some newspaper in the bottom of it and cover with a handful of freshly cut comfrey leaves.

4 Tip in the well-mixed potting mix and manure until the bucket is almost full to the brim (the level will settle and drop after watering).

5 Plant the individual pots of peas into the mixture. Five pots is about right for this size of container. Firm them in with your hands and, when all are planted, water thoroughly.

green*care*

Peas can dry out rapidly in containers, and they particularly dislike being baked by the sun. Keep them moist by watering in the morning, but do not overwater. As the first peas start to develop, a weekly liquid feed of seaweed is essential. When the peas start to ripen, pick regularly; this will encourage them to continue producing.

6 Finally, push the pea sticks (or canes) into place around the peas. Try to make the twigs all face inward. You can weave them together slightly to achieve this.

petit pois and sweet peas

Another cool-weather plant, the fragrant-flowered sweet pea makes an attractive companion for the vegetable pea and can be planted in the same container. There is no danger of mistaking the hairy, coarse pods of the sweet pea for the much more appetizing edible variety. Fortunately, since sweet peas make a lovely cut-flower display, they need to be picked on a regular basis to prolong their flowering season. They also need to be deadheaded regularly.

Above: *Sweet peas and edible peas make lovely companions; some varieties of edible pea also have colorful flowers, particularly the purple-flowered snow pea.*

Right: *The twining tendrils of the pea vine search out and quickly attach themselves to the twiggy branches of the pea sticks.*

Fava beans in a coconut sack

A good cool-climate alternative to lima beans, fava beans have long been popular in Britain, where they are called broad beans. They can be grown from zone 7 north as a spring crop. Given these conditions, they are easy to grow, and they taste delicious. Plant the seeds in early spring in seed-starting mix contained in cardboard tubes, such as those in the middle of toilet-tissue rolls or paper towels. Stand the tubes in a tray, and keep them under a cloche until the weather gets warmer.

Right: A synthetic vegetable sack, often used for packing and transporting potatoes and other root crops, makes an unusual and practical planter.

You can also grow the bean seedlings inside on a cool windowsill. The large seeds germinate quickly if the temperature is not too low. When the young plants are 4 inches (10cm) tall, they can be planted out in their permanent position, still in the cardboard tubes. The roots will happily penetrate the cardboard tubes, which will quickly biodegrade in the moist soil.

Below: The highly scented black and white flowers, which bloom all the way up the stem, are adored by bees. Young beans are forming at the base while the tops are still producing flowers.

Many vegetables are transported in the type of synthetic woven sack used here as a planting container—you could ask a greengrocer to save some for you. There is no need to put any drainage crocks at the base of the sack, because the close weave will allow moisture to escape freely. During hot weather, water the plants every day and take great care never to let the soil dry out.

greencare

The worst pests for fava beans are black aphids; but they can be warded off by pinching out the growing tips at the top of the stems. This will also encourage the pods to form.

Beans are hungry feeders and like rich soil, so incorporate some well-rotted manure into the multipurpose potting mix in a proportion of 3:1 mix to manure. The manure will also help the soil to retain water.

Borlotti beans
in a laundry basket

you will need

- **wicker laundry basket at least 28 inches (70cm) deep**
- **empty potting mix sack to go inside basket**
- **multipurpose loam-based potting mix**
- **well-rotted manure**
- **dibble**
- **borlotti beans ('Lingua di Fuoco 2')**
- **sheet of glass with rounded edges, such as a shelf**
- **rods or bamboo canes**
- **exterior house paint (blue or green)**
- **large-bristled brush**

Perhaps the most decorative vegetables, climbing beans have twining stems studded with flowers of many colors. Scarlet runner beans, with their abundant red flowers held on upright stems above lush green leaves, have long been valued for their appearance as well as their taste. Many more types are now available, with flowers ranging from white to pink, apricot, and orange. Why not grow a mixture of beans together up a bamboo tepee? Look in seed catalogs or garden centers and nurseries for varieties that can be sown at the same time.

The delicious and nutritious borlotti bean, a key ingredient of minestrone soup and other Italian dishes, has fat, cream-colored pods speckled with red blotches (hence their alternative name "cranberry bean"). Although it is not such a heavy cropper as other beans, its particular qualities make it indispensable for the vegetable garden. Unlike green and scarlet runner beans, borlotti beans are harvested from the pods before cooking. They are best eaten fresh, although they can be dried for winter use.

The sturdy cylindrical laundry basket used here is ideal. It has been painted with fence paint, which will help to prolong its life. Lining the basket with a strong plastic bag—in this case an empty potting mix bag—is essential to retain moisture. An added advantage, especially if you are growing vegetables on a balcony, is that the basket is much lighter than a conventional pot of a similar size.

Left: *As the beans ripen in late summer, the pods, originally cream speckled with red, turn a fiery crimson. To use them fresh, pick now and remove the nutritious beans from the pods.*

Right: *The sturdy bean plants twine tightly around the support rods or canes—they grow quickly at this stage.*

greencare

To grow borlotti beans, you will need a fairly deep, rich soil that has some well-rotted manure incorporated into it. Planting the beans in a deep container allows a lengthy root run and ensures that the soil will retain moisture. Keep the plants well watered, remembering to water in the evening on hot days. Like other leguminous (pealike) plants, borlotti beans are thirsty and hungry feeders. They will tolerate some shade, but need the warmth of the summer to ripen the pods.

1 Paint the basket following the instructions on the can. You may need to apply two coats to achieve complete coverage. Paint to a depth of 4 inches (10c) inside the rim. Allow to dry thoroughly.

2 Insert the plastic sack, having first cut a few slits in the base to allow for drainage. Roll the top edge over so that it fits snugly inside the basket rim.

3 Mix the potting mix and manure at a ratio of roughly 3:1. Throw a few crocks or stones into the basket to help with drainage before adding the mixture. Fill the last 8 inches (20cm) with multipurpose potting mix.

4 Make some holes with a dibble about 4 inches apart and sow one or two beans in each hole. You can remove the weaker of the two beans later.

5 Water the basket well, but make sure the water does not expose the planted beans.

6 Cover the beans with glass to raise the temperature of the soil. This will encourage the beans to sprout.

7 When the first true leaves have appeared, push a rod or cane into the earth beside it—the beans will quickly twine around it.

greencare

If you grow scarlet runner or green beans, pick them regularly to ensure a continuing crop. Allow borlotti beans to swell in the pod before harvesting. If you want to dry them, allow the pods to shrivel slightly on the vine before picking for storage.

fruits and berries

Even if you have only a small outdoor space, you can still grow some of your own fruit. Many fruit-bearing plants grow happily in containers, and some orchard trees have been bred especially for this purpose.

Grape vines thrive if grown in a warm, sunny place. They need a large container with enough soil to retain moisture, and should be pruned regularly to reduce leafy growth and promote a well-formed plant that will yield a fair harvest. Besides being productive, grape vines look wonderful when the leaves take on vibrant colors in the fall. A good nursery will advise on the best type of vine for your climate and conditions. In my opinion, the muscat-flavored grapes are the best.

Strawberries are perhaps the easiest fruit to grow in containers. Both cultivated strawberries and the wild or alpine type need to be kept moist and will tolerate some shade. Wild strawberries are a real treat and always a favorite with children. Allow the berries to ripen fully to a dark red color, and pick them to eat straight from the plant.

Many other kinds of berries, including blueberries, gooseberries, and currants, are suitable for growing in containers. Of these, blueberries are the clear favorite in the United States. Less familiar, but very popular in Europe, are currants—red, white, and black (not to be confused with tiny black raisins)—which appreciate cool summers and are well worth trying if you can offer them those conditions.

Physalis (also called cape gooseberries or tomatillos) are easy to grow and generally not attacked by pests or diseases. Seeds germinate quickly, or you can buy young plants. Like tomatoes, to which they are closely related, physalis need warmth and sun to ripen (berries will ripen indoors if left on a windowsill).

All fruit will benefit from a weekly high-potash liquid feed when the fruit begins to mature. A tomato feed is ideal; organic varieties are readily available.

Blueberries in a dolly tub

Above: *The small, attractive berries ripen from pale green to deep purple.*

Right: *Many birds love ripe blueberries. You can ward them off by loosely draping a piece of garden netting over the whole bush as the berries begin to turn blue.*

green*care*

Blueberries must be grown in ericaceous (slightly acid) potting mix, which is available, specially formulated, from garden centers. The soil should be kept moist, so it is best to stand the pot in a semi-shaded position. If you can use rain water for watering, it will help to maintain the acid conditions of the soil.

Blueberries, one of the sweetest fruits, are fairly easy to grow in pots. They look particularly good planted in any galvanized container, but they need plenty of space to thrive. With its pretty ribbed design and decorative rim, this old-fashioned English washtub, called a dolly tub, is ideal for the purpose. If you have only an ordinary washtub, that will do fine—the beautiful blueberries are the main attraction.

The blueberry plant grows into an attractive medium-size bush with bell-shaped, pale ivory flowers in late spring and beautiful autumn foliage. The berries, which grow in clusters, gradually ripen over a few weeks, turning from pale sage green to dark blue fruits covered in a bloom.

A fully mature bush will provide you with a daily handful of nutritious berries for a few weeks during the summer. If you have space, you can prolong the harvesting time by growing an early, a mid-season, and a late-cropping variety. Before buying a young plant from a nursery, note what the label says about cropping time—it would be a shame to buy a variety that crops only in midsummer while you are away on vacation.

Strawberries in wooden fruit boxes

you will need

- old wooden fruit or wine boxes
- plastic woven potato sack for lining
- sharp sand
- potting mix
- well-rotted manure
- large staple gun

Strawberries are very well suited to being cultivated in containers, but to achieve a successful crop you need to follow a few basic rules. Choose a humus-rich, moisture-retaining potting mix, and mix in bonemeal and well-rotted manure before planting. To cut down on water loss, line your container with woven plastic sack fabric. Strawberries, particularly the delicious alpine variety, can tolerate a little shade.

Regular feeding with a seaweed extract will ensure large and healthy fruits. Straw placed around the ripening plants will keep them clean, and its light color will raise the temperature around the fruit by reflecting the light. You may have to cover the fruits with net as they ripen to protect them from birds.

1 Cut the potato sack into two pieces. Place one piece inside the box, folding over the edges, then staple in place. Use the other piece to cover the rest of the wood.

2 Put a layer of sharp sand over the base to improve drainage.

Above left: *Once the flowers appear, make sure the box is near other flowering plants to encourage bees to visit.*

Right: *An old wooden fruit box, stamped on the slatted sides with the name of the grower, makes an attractive and practical container in which to grow strawberries.*

green*care*

Some gardeners put the strawberries into a glass jar when they are green, creating a mini greenhouse to speed up ripening. This also has the advantage of protecting them from birds.

3 Add a generous layer of potting mix to the box, covering the sandy base.

4 Add a layer of manure and mix well. Mix in more potting mix and some bonemeal.

5 Plant the strawberry plants equidistantly from one another in the box, pressing the potting mix down firmly around them.

6 The plants will grow quickly. If you keep them well watered, they will produce many flowers. Watch for pollinating insects at this point. When the strawberries begin to set, place some clean straw underneath and around them, to lift them off the soil.

potting strawberry runners

Strawberries are very obliging plants. After flowering, they start to send out long stems, known as runners, from the crown of the parent plant. If potted, a runner produces roots and becomes a new healthy plant, ready to produce fruit for the next season.

Above: *The six plants cultivated in this wooden box could produce at least twenty new plantlets.*

Right: *Fill a pot with loam-based potting mix. Place a plantlet in the center of the pot, and "pin" it in place with a hairpin-shaped piece of strong garden wire. Leave this plantlet attached to the parent plant until fully rooted, at which point it can be detached. Plant out in the fall in a new container.*

Alpine strawberries

Plant alpine strawberries in the same way as that described for dessert strawberries on pages 128–131. They will ripen earlier and go on fruiting for longer than their larger counterparts. The tiny pointed fruits are produced on high stems, often above the leaves. Curiously, birds don't seem particularly interested in the alpine variety.

Right and far right: *Alpine strawberries are easy to grow and delicious. Luckily, they go on producing fruit for the whole summer. They are happy to grow in part shade, as they originate in clearings and on the edges of alpine forests.*

A tub of physalis

The many common names for *Physalis*, which include tomatillo and ground cherry, suggest its versatility. Originally from South America and now grown commercially in South Africa (hence another name, Cape gooseberry), it can be used in both sweet and savory dishes.

The flowers are exquisite and unusual, hanging timidly down from the leaf axils. As they grow, they develop into delicate papery lanterns, which enclose the deep yellow berries. When ripe, they have a delicious musky taste. In Mexican cooking they are used in salsa, but they can also be used as a garnish on cakes and in fruit salads; they make a wonderful jam if you can harvest enough of them.

Physalis is a vigorous, easy-to-grow plant, which will happily self-seed in the right conditions, but it dislikes a soil that is too rich. Although it can be sown from seed early in the year, you may find it easier or more convenient to buy ready-grown young plants at the right time to plant outside. You will find them for sale alongside young tomato plants in early summer. The fruits ripen late in the summer; to see if they are sweet and ready to eat, pick one and try it. They will ripen from the base of the plant upward (somewhat like tomatoes) and will continue to crop over a few weeks.

Above left: *The pale yellow flowers, with chocolate-spot centers, are unique and intriguing. They appear evenly all over the plant where the leaves attach to the stem.*

Above right: *Each fruit is enclosed by a papery lantern, which begins a pale green and gradually turns to a straw yellow as the enticing yellow fruit within ripens.*

Right: **Physalis** *grows well in containers. Its fruit makes an unusual and refreshing addition to salads and desserts, and can also be made into jams and jellies.*

green*care*

Since *Physalis* plants are easy to grow, they will adapt to a variety of containers, so long as the pots are deep enough to contain the roots and hold sufficient water. Besides being a pretty shape and color, this enameled washtub is ideal; any metal container, being nonporous, has the advantage of conserving moisture in the soil. Punch a few holes in the base of the bowl, and add a layer of gravel or clay pot crocks to promote good drainage. Finish with a hard mulch of stones, china shards, or shells to prevent unnecessary moisture loss from the surface of the soil.

Pot-grown grape vine

Perhaps surprisingly for such vigorous plants, grape vines will thrive in containers, provided they're large ones. To produce worthwhile fruit, the vines need warmth and sun. A south- or southwest-facing wall is ideal, providing shelter and radiated heat from daytime sunshine. Grape vines are decorative climbing plants with deeply cut leaves, which provide a

mellow show as the bunches of black or white grapes ripen in late summer. Look for an outdoor type that suits your location. Good garden centers will offer a small variety of vines, but for more specialist advice, consult a dedicated grower.

Grape vines need a rich loam-based potting mix. When the grapes have matured, feed the plant weekly with a high-potash feed—liquid comfrey (see page 91) or tomato food is fine. Regular pruning and training are essential. In early summer, prune the young green shoots, known as laterals, to two leaves beyond the flowers or young grapes. In winter, prune back severely to the wood, a short distance beyond the previous year's growth. It's worth consulting a book or magazine with step-by-step instructions for pruning—when you've done it once, you'll know what to do each year.

You will be able to grow enough grapes to have a good few bunches to eat as fruit. Alternatively, you could make a small quantity of wine, grape juice, or jelly. Even the leaves are edible (after being pickled in brine), being widely used in Greek cooking to make little parcels stuffed with rice and ground meat.

greencare

A grape vine needs a large container—one that can hold enough potting mix to feed the roots and to make a firm anchor to stop the plant from becoming top-heavy and pulling the pot over. The clay pot used here is a simple version of the traditional terracotta pot in which grape vines are so often grown.

If you want to train a grape vine up an obelisk or grow it on a frame of canes, you need to tie in the laterals repeatedly until you have the size and shape of vine that you want.

Above left: *Given the right conditions, grapes will thrive in containers.*

Right: *This gray clay planter attractively complements the ripe black grapes it contains. A vine is happiest sited in the sunniest position in the garden. Mulch the surface of the soil with an assortment of decorative pebbles.*

Red and white currants

Right: *The glistening berries of the red currant are the jewels of the fruit garden. One bush can produce enough berries to make a few jars of delicious jam or jelly.*

Far right: *White currants are heavy croppers. Curiously, they are sweeter to taste than the ruby red berries of the red currant. The berries can be picked all at once when they are all fully ripe. Test for ripeness by tasting.*

greencare

All currants can grow well in large containers, but they need to be kept moist and should not be in relentless full sun. Since currant bushes are perennials, you will need to refresh, feed, and replenish the soil each spring.

Currants come in three types: black currants, red currants, and white currants. These tender berries really are worth growing yourself, since it is unusual to find them in stores—they are laborious to pick and don't travel well. Black currants, the largest of the three, are a deep purple-black; they are tart and best eaten cooked, making excellent jam, pies, and puddings. The smaller red currants, with their glistening vermilion berries, grow in delicate bunches called strings. They are much sought after by garden birds and need to be covered with nylon mesh. The similar but sweeter white currants generally escape the ravages of the birds, which are not attracted by their pale translucent berries.

Red currants and white currants are delicious eaten fresh. Both of these varieties also make lovely clear jelly; add some chopped homegrown red chili to make an excellent accompaniment to meat and poultry dishes.

Useful addresses

Backyardgardener, LLC
P.O. Box 23598
Federal Way
WA 98093-0598
www.backyardgardener.com
Information site with forums
and store providing products
for organic gardening.

City Gardening magazine
This is a quarterly magazine
written for the gardening hobbyist
(non-professional) and gardening
professional who grow and
cultivate flowers, fruits, herbs,
shrubs, trees, and vegetables in
a city. Subscribe via the website:
www.citygardeningmagazine.com

GardenMandy
www.gardenmandy.com
Organic gardening online tips,
tricks, and information.

Garden Organic
(previously HDRA)
www.gardenorganic.org.uk
A British website, but at the
forefront of organic gardening
and research. Lots of useful
information.

LandscapeUSA
13126 NE Airport Way
Portland
OR 97230
tel: 800.966.1033
customer service email: sales-
landscapeusa@landscapeusa.com
www.landscapeusa.com

www.allotment.org.uk
Useful information on vegetable,
fruit, and herb gardening,
especially on small plots or in
community gardens.

www.gardensimply.com
Monthly tips on sustainable
organic gardening practice.

www.gardening-advice.org
Gardening tips and advice.

www.pennystomatoes.com
Tomato seeds and recipes.

Seed suppliers

Eden Organic Nursery Services Inc.
Contact address:
P.O. Box 4604
Hallandale
FL 33008
tel: (954) 382-8281
info@eonseed.com
Shop address:
2021 SW 70th Ave
#B9 Davie
FL 33317
www.eonseed.com

Heirloom Seeds
PO Box 245
W. Elizabeth PA 15088-0245
mail@ heirloomseeds.com
tel: (412) 384-0852
www.heirloomseeds.com

Johnny's Selected Seeds
955 Benton Avenue
Winslow
Maine 04901
www.johnnyseeds.com
Toll Free: 1-877-Johnnys
(1-877-564-6697)
Business Line: 1-207-861-3900
Catalog Store: 1-207-861-3999

Marianna's Heirloom Tomatoes
1955 CCC Road
Dickson
TN 37055
615/446-9191
www.mariseeds.com
mj@mariseeds.com

Neseed.com
3580 Main Street
Hartford
CT 06120
Toll free phone: 800-825-5477
sales@neseed.com

Park Seed Company
1 Parkton Ave
Greenwood
SC 29647
Orders: info@parkscs.com
For gardening advice:
gardener@parkscs.com
www.parkseed.com

Pepper Joe's, Inc.
7 Tyburn Court
Timonium
MD 21093
pepperjoe@comcast.net

Richters Herbs
357 Highway 47
Goodwood
ON L0C 1A0
Canada
Tel: +1.905.640.6677
www.richters.com

Seeds from Italy
PO Box 149
Winchester
MA 01890
Email:seeds@growitalian.com
or bmckay@growitalian.com
tel: 781 721 5904
www.growitalian.com

Seeds of Change
www.seedsofchange.com
1-888-762-7333 to speak with a
customer service representative
about your order
Store-finder: 1-888-762-4240

Territorial Seed Company
20 Palmer Ave
Cottage Grove
OR 97424
Phone Orders: 800-626-0866
Customer Service/Gardening
Questions: 541-942-9547
info@territorialseed.com

Victory Seed Company
P.O. Box 192
Molalla
Oregon 97038
tel: (503) 829-3126
info@victoryseeds.com
www.victoryseeds.com

Index